"*To learn to classify is in itself an education*" — ALEX. BAIN

ABRIDGED

Decimal Classification

and

Relativ Index

for libraries, clippings, notes, etc.

Edition 2

By

Melvil Dewey MA LLD

Forest Press
Lake Placid Club N Y
1915

CONTENTS

	PAGE
Explanation	3-24
Change to full form	3
Plan	3
Summaries	4
Tables	4
Choice and arrangement of heads	5
Sequence of allied subjects	5
Coordination	5
Catch titles	5
Form distinctions	6
Mnemonics	6
Relativ subject index	7
What the relativ index includes	8
Relativ location	8
Sizes on shelves	10
Catalogs	10-11
Name catalog	10
Shelf-list	10
Subject catalog	11
Dictionary catalog	11
Suggestions to users	11-17
Numeration	11
Number of figures used in class numbers	11
Suggestions (continued)	
Familiarity with tables	12
Use of index	13
How to determin subject of book	13
Assigning class numbers	14
Book numbers	17
Variations practicable	18-22
Cautions	18
Letter notations for changes	19
Fiction and juvenils	19
Biografy	20
Parallel libraries	20
Combining language and literture	21
Broken order	21
Pro and con division of topics	22
Unassigned numbers	22
Other uses	23
Bookstores	23
Scrap-books	23
Index rerums	23
Topical indexes	24
Tables	25-74
Index	75-192
Index supplement	193

Publisher's Note

The simplified spellings used are strongly recommended for general adoption by both the American and English Philological Associations, including nearly all the prominent scholars in English now living. The publishers regret the prejudis that certain readers will feel against these changes, but after careful study of all objections urged against them, they find the weight of scholarship and reason wholly in their favor, and feel compeld to bear a share of the prejudis which some must endure before the great benefits of a rational orthografy can be secured.

Full information on request from Simplified Spelling Board, 1 Madison av., New York.

Abridged Decimal Classification[a]

EXPLANATION

Before beginning to use the classification at least the paragrafs in large type should be carefully read. Those interested will find still fuller explanation in the smaller type.

This abridgment is made in answer to a strong demand for a short form adapted to the needs of small and slowly growing libraries. They wish short class numbers, but it is a mistake to assume that they will have only works on general subjects. Even the smallest library is likely to have a few books or pamflets on the most specific topics of the omitted subheads, and as these minor subjects will differ in each library, it is obviously impossible to make any abridged classification that will meet every want. Cases will certainly arise that can be met only by the unabridged classification, which aims to include all subjects treated in books, pamflets, or articles.

(See also Relativ subject index, p. 7; Use of index, p. 13.)

Change from short to full form. These short (three-figure) forms can be changed to the full class numbers at any time without other alteration than adding extra figures to those here given.

Plan. In this classification the field of knowledge is divided in nine main classes, numberd 1 to 9. Cyclopedias, periodicals, etc. so general in character as to belong to *no* one of these classes, are markt 0 (naught), and form a tenth class. Each class is similarly separated into nine divisions,

---

a The confusion and annoyance to the many users of this system caused by printing unauthorized variations have forced the publishers to insist strictly on the protection afforded by the copyright laws. Every library and individual user has, however, entire freedom to make such variations as he thinks he needs, under the simple restrictions found necessary to protect the rights of others as explaind on page 19, Letter notations for changes.

general works belonging to *no* division having 0 in place of the division number. Divisions are similarly divided into nine sections, and the process is repeated as often as necessary, but in this abridgment 0 is not used beyond the third figure. Thus 512 means Class 5 (Natural Science), Division 1 (Mathematics), Section 2 (Algebra), and every Algebra is numberd 512. This class number, giving class, division, section and sub-section, if any, is applied to every book and pamflet belonging to the library.

Where 0 occurs in a class number, it has its normal zero value. Thus a book numberd 510 is class 5, division 1, but *no* section; i. e. the book treats of division 51 (Mathematics) in general, and is limited to no one section, while geometry which is so limited is markt 513; 500 indicates a treatise on science in general, limited to no division. A naught occurring in the first place would in the same way show that the book is limited to no class; e. g. a general cyclopedia which treats of all nine classes.

The books are arranged on the shelves in simple numerical order, all class numbers being decimal. Since each subject has a definit number, all books on any subject must stand together. These tables show the order in which subjects follow one another. Thus 512 Algebra precedes 513 Geometry, and follows 511 Arithmetic.

Summaries. The first page of tables shows the 10 Classes into which all topics are divided. The next page shows the nine Divisions of each of the 10 classes, and is useful as a bird's-eye view of the whole scheme.

Tables. Following these two summaries is the abridged classification, which repeats, in proper order, all the classes, divisions, and sections, with such sub-sections as are used in this abridgment. Synonymous terms, examples, brief notes, dates, and various catch-words are often added to the simple heads for convenience of users, who thus get a fuller and clearer idea of the field which each number covers. Therefore all references to numbers should be lookt up in the tables; never in the summaries, which are really only a table of contents of the complete tables.

Choice and arrangement of heads. The selection and arrangement of heads can not be explaind in detail for want of space. In all the work, philosofic theory and accuracy have been made to yield to practical usefulness. The impossibility of making a satisfactory classification of all knowledge as preservd in books, has been appreciated from the first, and theoretic harmony and exactness have been repeatedly sacrificed to practical requirements.

Sequence of allied subjects. Wherever practicable, heads have been so arranged that each subject is preceded and followd by the most nearly allied subjects, and thus added convenience is secured both in catalogs and on shelves; e. g. Building (690) follows Mechanic trades (680) at the end of Useful arts, and Architecture follows near the beginning of Fine arts.

The student of Biology (570) finds fossil life or Paleontology (560) before, and vegetable life or Botany (580) after, this followd in turn by animal life or Zoology (590), ending with Mammals (599); while Useful arts (600) begin with human Anatomy (611) under Medicin, thus giving a regular progression from the fossil plant thru the vegetable and animal kingdoms to the living man.

Coordination. Theoretically, the division of every subject into just 10 heads is absurd. Practically, it is desirable that classification be as minute as possible, without use of added figures; and the decimal principle, on which our scheme hinges, allows 10 divisions as readily as less.

This principle has proved wholly satisfactory in practice, tho apparently destroying proper coordination in some places. In the full classification this difficulty is entirely obviated by the use of another figure, giving nine sub-sections to any subject of sufficient importance to warrant closer subdivision. In history, where classification is made chiefly by countries, a single figure is added, as in the other classes, to give a tenfold geografic division. Wherever naught followd by another figure is added, it indicates a division into periods.

As all history is by countries, and as close geografic subdivisions are needed for local history and various other uses, the rule is whenever the same number is subdivided both geografically and by periods, to insert naught before the time or period figure; the 0 showing a change from geografic to period division, e. g. 942.05 England in time of the Tudors.

As in every scheme, many minor subjects are under general heads to which they do not strictly belong. In some cases, these heads are printed in a distinctiv type; e. g. 429 Anglo-Saxon, under English philology. The rule has been to assign these subjects to the most nearly allied heads, or where it was thought they would be most useful.

Catch-titles. In naming heads, strict accuracy has often been sacrificed to brevity, for short familiar titles seemd more important than that the heads given should express with fulness and exactness the character of all books catalogd under them.

Form distinctions. The classification is mainly by *subject* or *content* regardless of form; but a form distinction for general treatises, which is found practically useful, is introduced at the beginning of the class, where the first numerals would otherwise be unused. These form distinctions apply only to general treatises, which, without them, would have a class number ending in two naughts.

Thus, in Science there are many compends, dictionaries, essays, periodicals, and societies, treating of Science in general, and so having o for the division figure, but treating it under different forms, and, therefore, divided into sections according to this form: 501 for philosofy or theories of Science, 502 for compends, 503 for dictionaries. This treatment is as nearly as practicable uniform in all classes. Brewer's *Reader's handbook* is 803, the first figure being 8 because the book is clearly on literature; the second figure o, because limited to no division of class 8; and the third figure 3, because the book is a dictionary.

Mnemonics. Arrangement of heads has been sometimes modified to secure mnemonic aid in numbering and finding books without the index; thus China has always the number 1. In Ancient history, it has the first section, 931; in Modern history, under Asia, it has 951. Similarly the Indian number is 4; Egyptian, 2; English, 2; German, 3; French, 4; Italian, 5; Spanish 6; European, 4; Asian, 5; African, 6; North American, 7; South American, 8; and so for all divisions by languages or countries. Italian 5, for instance, is in 035, 055, 065, 450, 850, 945, and many others. This mnemonic principle is specially prominent in Philology and Literature, and their divisions, and in *form* distinctions used in the first nine sections of each class. Philosofy, methods, or theory occurring as a head is always 1; dictionaries and cyclopedias, 3; essays, 4; periodicals, 5; associations, institutions, and societies, 6; education, 7; polygrafy or collections, 8; history, 9. In numerous cases several minor heads are groupt together as Other, always numberd 9.

While Italian is always 5, 5 is by no means always Italian. Grammar is 5, Periodicals are 5, Asia is 5, Oratory is 5, etc. Even were it possible, to limit 5 to Italian would waste numbering material, and results would not justify cost. The purpose is to give practical aid, not to follow a fanciful theory. A cataloger marking a German grammar, remembers that all Philology begins with 4, and, as German is always 3 and grammar 5, he knows the number must be 435. Italian (5), poetry (1), is as plainly 851 with no danger of being mistaken for "poetry of grammar" or "theory of Asia," because the numbers also have that meaning. This feature is an *aid*, not the regular method, and in all doubtful cases one refers at once to index or tables. Suggested difficulties are usually creations of ingenious theorists and not an outgrowth of practical experience in using this plan.

Wherever practicable, this mnemonic principle is used in subdividing sections. 558, Geology of South America, is subdivided by adding the *sections* of 980, History of South America. Geology of Brazil then must be 558.1: mnemonicly the first 5 means Science; the second 5, Geology; 8, South America; 1, Brazil. Any library attendant or regular user of the scheme recognizes 558.1 at a glance as Geology of Brazil. This mnemonic feature is of great practical utility in numbering and finding books without catalog or index, and in determining the character of any book simply from its call number. The advantage of mnemonic correspondence has however never outweighd any claim of greater usefulness. In many cases choice between numbers was hardly perceptible; e. g. whether in Philology the order should be French, Spanish, Italian, or French, Italian, Spanish. In such cases, preference was given to mnemonic numbers, and 20 years' experience has now proved this wisest.

Relativ subject index. Following the tables is the most important feature of the system, an alfabetic index of all the heads, which refers by class number to the exact place of each in the preceding tables.

The index might be made on any of the various dictionary plans. The simplest was however chosen. Only short heads are given with a brief indication in doubtful cases of the point of view taken in assigning the class number.

This index is designd to guide both in numbering and in finding books. In assigning numbers, the most specific head that will contain the book having been determind, reference to that head in the Index gives the class number which should be assignd. Conversely, in finding books on any given subject, reference to the Index gives the number under which they are to be sought on the shelves, in the shelf list, or in the subject catalog. Whenever a subject not in this index comes up, it should be lookt up in the full index so that the classification may be uniform with that of other users.

The index aims to give similar or synonymous words, and the same words in different connections, so that any person of intelligence will

hardly fail to get the right number. A reader wishing to know something of the tariff looks under T, and, at a glance, finds 337 as its class-number. This guides him to the shelves, to all the books and pamflets, to the shelf list, to the subject catalog on cards, to the index of books charged at the loan desk, and, in short, in simple numeric order, thruout the whole library to anything bearing on the subject. If he turns to the tables, he will see that it means class 3, Sociology; division 3, Political economy; section 7, Protection and free trade; but the number alone is enough to classify the book or to find it, for either cataloger or reader. If he had lookt under P for protection, or F for free trade, or D for duties, or C for customs, or under any other leading word relating to his subject, he would still have been referd to 337.

Had he lookt for "railroad" he would have found after it 13 separate entries, each preceded by a catchword indicating the phase of the subject in the scheme. A book on railroads may be a treatise on the desirability of government ownership, control, etc., and then is clearly a question of social science; or it may be a practical handbook for an employee explaining business methods of railroading, running trains, handling freight, etc., when it is as clearly one of the useful arts. The cataloger knows to which of these heads his book belongs, and the reader knows which phase of the subject he wishes to examin. Moreover, the 3 and 6 beginning the numbers indicate clearly the character of each class. But even if the significance of these figures were entirely disregarded, no confusion would result; for, on consulting either of the numbers in the catalog, in the tables or on the shelves, the difference would be clearly seen.

What the relativ index includes. This index includes also, as far as found, all synonyms or alternativ names for the heads, and many other entries that seemd likely to help a reader find his subject readily. Tho the user knows just where to turn to his subject in the tables, if he consults the index he may be sent also to other allied subjects; where he will find valuable matter which he would otherwise overlook.

Most names of countries, towns, animals, plants, etc. have been omitted, the Index containing only those which are in the tables; e. g. it can not enumerate all the species of trilobites, but when the classifier has found from the proper reference books that Remopleurides is a trilobite, the index will send him to 565 and he can classify his monograf on that subject.

Relativ location. Economy and simplicity calld not only for the subject index, but also for some plan of consolidating the two sets of marks heretofore used; the one telling

of what subject a book treats, the other where the book was shelvd. By relativ location and decimal class numbers, the simple arabic numbers have been made to tell of each book and pamflet, both *what* it is, and *where* it is.

In arranging books on the shelves, absolute location by shelf and book number is wholly abandond, relativ location by class and book number being one of the most valuable features of the plan.[1] In finding books, the class and book numbers on the backs are followd, the upper being the class and the lower the book number. (*See* p. 17, Book numbers.) The class is found in its numeric order among the classes, just as the shelf is found in the ordinary system. Shelves are not numberd, as increase of different departments, opening of new rooms, and any arrangement of classes to bring books most circulated nearest the delivery desk, will at different times bring different class numbers on any given shelf. New books, as receivd, are numberd and put in place, in the same way that new titles are added to the card catalog.

Not only do all the books on any given subject stand together, no additions or changes ever separating them, but the most nearly allied subjects precede and follow, they in turn being preceded and followd by other allied subjects, as far as practicable. Readers not having access to the shelves find short titles arranged in the same order in the shelf list, and full titles, imprints, subject and cross references, notes, etc. in the subject catalog.

Parts of sets, and books on the same or allied subjects, are never separated as they are sure to be, sooner or later, in every library arranged on the old plan, unless the great and needless expense of frequent rearrangement and recataloging be incurd. In this system both class and book numbers remain unchanged thru all changes of shelving, buildings, or arrangement.

Among hundreds of points raisd by librarians, as to its practical workings and usefulness, the only one in which it is

---

[1]For full explanations regarding forms of book numbers best used with this scheme of classification, see p. 17, and *Library notes* no. 11.

not shown to be equal or superior to the old systems is that in this relativ location a book which this year stands, for instance, at the end of a certain shelf, may not be on that shelf at all another year, because of uneven growth of the parts of the library. This slight objection, however, inheres in any system where books are arranged by subjects, rather than by shelves, windows, doors, and similar non-intellectual distinctions.

Sizes on shelves. Most libraries have abandond close distinction of sizes. It is true that this distinction saves a little space, but at far too great a cost; for every distinction of sizes makes a parallel classification. If books are groupt in five sizes, one must look in five places before he can be sure of having seen all the books on a given subject.

It is better to shelve octavos and all smaller books together in one series, and arrange in parallel libraries only quartos and folios, which are too large to stand on the regular shelves, showing the series in which any oversize book is put, by a size letter prefixt to the book number; e. g. 749 qB, to show that book B on Artistic furniture is too large for the regular shelves, and so is placed in the "q" or quarto series. Another way is to use a wood or pasteboard dummy, to show the location of a book not in its regular place. But, however solvd, there is no conflict between the size problem and the *Decimal classification.* (See also *Library notes* no. 11, vol. 3, p. 428–29.)

CATALOGS

Any system of catalogs may be used with this scheme. But the two essentials of even the simplest system are the author (or preferably name) catalog and the shelf list. The chief use of the system for catalogs is for the shelf list and for the clast catalog on cards.

Name catalog. In the name catalog on cards, arranged strictly by names of authors and of persons or places written about, the class number holds a subordinate place, yet is constantly useful. If printed, it appears in a single colum as in the relativ index, and, where there is no subject catalog, one can rapidly pick out the books on any desired topic by glancing down the colum for the class number wanted.

Shelf list. The class number makes the list the most useful form of brief subject catalog. Each page gives in colums accession number,

author's name, and brief title of every book in the library on the specific subject bearing that class number.

Subject catalog. In the card catalog of subjects the classification is mapt out above the cards by projecting guides, making reference almost instantaneous. Subjects are arranged in the 1, 2, 3 order of their decimal subject numbers exactly as in the classification tables, and the cards of each subject are then further arranged alfabetically by authors.

The printed subject catalog on this plan is also the most compact and satisfactory in use. Under each class number are given the resources of the library on that subject, the heading giving for convenience the name as well as the number of the subject; e. g. "513 Geometry." General notes are printed in finer type under general heads, and an index at the end shows where to open the book to find any topic. As class numbers are put in place of page numbers, this index serves for any catalog, list or library arranged on this plan.

Dictionary catalog. A dictionary card catalog may be as readily used with this system as with any other, but its advantages are largely supplied by the subject index, which is in itself a skeleton dictionary catalog. If however the subject card catalog is clast, the alfabetic index referring to class numbers makes the single set of cards serve both as clast and dictionary catalog. Instead of giving the book titles under each head, the index number refers to all those titles simply and directly.

The advantages of the dictionary and clast catalogs are therefore united, not by mingling them together, and so losing much of the simplicity of one and as much of the excellence of the other, but by really using both, each with its own merits.

SUGGESTIONS TO USERS

In referring to tables, hold the book in the right hand and turn with the left. The class numbers then show very plainly on the left margin and reference is greatly expedited. Some prefer to hold the book in the left hand, but in any case the eye should follow the left margins wholly.

Numeration. In thinking or speaking of the class numbers, to avoid confusion *always divide at the decimal point, and name it;* e. g. read 942.9, "nine forty-two, point nine," never "ninety-four twenty-nine." If "point" were omitted the ear might readily interpret 270.2 as 272, while "two seventy, point two" can never be misunderstood.

Number of figures used in class number. In very small collections two figures might do till growth required further

division; but it is economy, and saves handling the books a second time to use at least three figures at first, even in the smallest collection.

Whether there are one or 1,000 books on any topic, they take no more space on the shelves if clast minutely, and the work is done once for all. When large accessions come, even if a century later, this number will not have to be alterd. A library having only 20 books on economics should use the full scheme, for the whole 20 would go on a single shelf, and take no more room, and the index would refer more accurately to what was wanted. The number of books you have on any subject has in this system no special weight. In a relativ location, any number of consecutiv topics without a book yet in them, waste no space on shelves or in catalogs; the numbers are merely skipt. This plan not only does no harm, but has the considerable negativ value that looking for the number and finding it blank or skipt, shows that you have nothing on the subject, —a piece of information second in value only to finding something, for one need no longer search.

The practical objection to close classing is that it gives a longer number to charge in a lending library. In a reference library full sub-sections should be used. Where a short number is imperativ, it is well to give the full class number on another part of the book plate, not to be used in charging, but as a guide to the contents of the book. Thus when a classifier has once examind a book, and found out just what it is about, the record is preservd for the benefit of others.

Familiarity with tables. Get a general knowledge of the scheme by learning the 10 main classes (you will soon know the 100 divisions also without special study), so that you can tell to what subject a given number belongs from its first figure without referring to the tables. Specific knowledge of subdivisions will come gradually, but rapidly, from use. Using the tables alone, and then always verifying your result by the index, you will more rapidly acquire knowledge of the classification and facility in its use. To do this decide first to which of the 10 classes the subject belongs; next, take that class as if there were no other, and decide to which of its 10 divisions the subject belongs; then, in the same way, select section and sub-section, thus running down your topic in its grooves, which become tenfold narrower at each step.

If there are only 10 books on a given subject it is useful to have them still farther groupt by topics or form of treatment, for otherwise, they have only accidental order which is of service to no one. If a reader wishes a specific subject, he is sent instantly to the exact place by the subject index; if, however, he wishes a specific book, he should go, not to the shelves, but to the name catalog, where he can find its place quickest. If he wishes to study the library's resources at the shelves, he will be *greatly helpt by close classing.*

A library recently adopted the plan of putting all books of a division together, if they had but few; e. g. all mathematical works were markt 510. It took just as many figures, cost just as much labor in most cases, and, if a man wanted the one calculus in the whole library, he had to search through the 150 volumes in 510, when otherwise he would instantly have found it standing alone as 517.

A teacher showing his pupils the material on any subject, if there are only 20 books, would surely put together those covering the same points, even if there were but two. Much more should librarians group closely their greater collections, that readers may gain something of the advantages of an experienced guide.

This gives every specialist his own distinct library. If a student of science in general, he is sent to class 5; if his department be zoology, his library is 59; if his specialty is shells, he finds all works and references on that subject in library 594. Whether a specialist needs it or not, every subject being a library by itself shows resources and wants as no catalog can show them. A catalog can not be made to take satisfactorily the place of handling books themselves. This advantage weighs most in a college or society library, where many go to the shelves; but even if librarians only are admitted, close classification is worth its cost in the added power it gives.

5 The predominant tendency or obvious purpose of a book usually decides its class number at once. Still a book often treats of two or more different subjects. In such cases assign it to the place where it will be most useful, and make references under all subordinate subjects. Give these reference numbers both on book plate and subject card, as well as on reference cards.

It is one of the markt advantages of the plan that these references, notes, etc., may be added from time to time as found convenient. It is necessary at first to find only the predominating tendency of the book in order to classify it. Subject references are added whenever found necessary.

Add these numbers indicating more closely the character of the book as rapidly as possible, and invite all specialists to call the librarian's attention to every desirable subject reference noticed in their reading. These numbers take little room, are easily added, and in most cases are very valuable.

6 If two subjects have distinct page limitations, class under the first, and make analytic reference under the second. But if the second is decidedly more important or much greater in bulk, class under that, with reference under the first. Always put a book under the first subject, unless there is good reason for entering it under another.

7 Consider not only the scope and tendency of each book, but also the nature and specialties of each library.

Any subject of which a library makes a specialty naturally "attracts" allied subjects. This influence is strongest in minute classification. To admit this variation, many subjects in the present scheme have two or more places, according to these different sides; e. g. a book on school hygiene, which a medical library puts under 613, has also a place in 371, where the educational specialist requires it.

8 If a book treats of a majority of the sections of any division, give it the division number, instead of the most important section number with subject references. Unless some one section is so prominently treated as to warrant placing the book in it, class a book on four [illegible] sections under the division number; e. g. cla[illegible] on light, heat, and sound, under the h[illegible] and refer to it from the othe[illegible] class as 530, or gene[illegible] drosta- tics [illegible]

This last practice constantly grows in favor, and many librarians now largely disregard uniform bindings and "series" lettering, and, unless contents of volumes are so connected that they can not be separated, class each under the most specific head that will contain it.

11 Class translations, reviews, keys, analyses, answers, and other books about specific books with the original book, as being there most useful.

Book numbers. With this abridged classification the following rough alfabetic arrangement of books in each class by author's surnames is recommended:

Keep together names beginning with the same letter by marking books by the first author under any letter with the initial of the author's surname; books by the second author under that letter with the author's initial followd by 1. Use the same initial and figure for all books by the same author in any one class, and distinguish different books by adding lower-case letters. E. g. if in class 942, History of England, the first books under G were Green's *History of the English people*, Gardiner's *Outline of English history*, Guizot's *History of England* and Green's *Short history of the English people*, and were receivd in the order named, the book number of Green's *History* would be G, Gardiner's *Outline* would be G1, Guizot's *History* G2, and of Green's *Short history* Ga.

In order to keep lives of the same person together book numbers in individual biografy should be assignd from the name of the person written about instead of from name of writer. These biografies are markt B and arranged in a single alfabetic series.

For fiction, omit the class number altogether and alfabet by authors' surnames. Both fiction and biografy should be numberd by the author table devised by C: A. Cutter, and different books by the same author distinguisht by adding a, b, c, etc. to the Cutter number.

For a full discussion of book numbers see *Library notes* no. 11, which is wholly given up to an explanation and tables of the systems which have been found most useful.

VARIATIONS PRACTICABLE IN ADJUSTING TO SPECIAL LOCAL REQUIREMENTS

Some users assume that adoption of the *Decimal classification and relativ index* carries with it other parts of the general system which the author has used at Amherst, Wellesley, or Columbia colleges or in the New York state library. In fact, the plan in each of these libraries differs somewhat from all the others, and many of the more than 300 libraries now using this system have adopted still other variations ; for the special constituency, circumstances, and resources of each library must be considerd in deciding what is best for it. This decision should be made by one familiar, not only with the library and its needs, but also with all methods of any merit and the comparativ ease and cost of introducing them into any given library.

Cautions. Having decided to adopt the system in its decimal form as workt out and printed, it must be determind whether to adopt certain variations, noted in 1–4 below as practicable, and in some cases useful and desirable. The inexperienced user is very likely to feel entirely competent, after once reading the tables (in fact, without reading more than a single page regardless of its bearings on hundreds of other places, and without so much as looking at the author's explanations), to institute a series of "improvements." Experience proves that nothing could be more disastrous. It seems a simple matter to put a topic a line higher or lower, but in some cases this may affect over 100 different entries in the index, and there is no possible way to be sure of correcting them except by examining each of thousands of heads. Any changes except as authorized under the next heads, would conflict with the full tables, so that closer classification, if required later by the growth of the library, could be made only by changing whole numbers instead of by simply adding extra figures. Care has been taken that no number in this short form should vary from that in the full tables ; e. g. instead of printing as the number for Ireland only 941.5, the numbers 941.6–.9 are given for the different groups of counties, just as in the full tables. If the numbers for these counties were designated by adding figures to 941.5, the closer classification would be omitted in this edition, for small libraries are not likely to have enough on Ireland to make subdivision necessary. But if the books on the various Irish counties were all numberd 941.5 the number would disagree with those on all printed catalog cards and cooperativ lists and, if the change to full class numbers should ever be made, the last figure on every book and on every record of that book would have to be alterd, while nothing would have been gaind originally in brevity.

Frequently proposed changes, carefully studied out and submitted as improvements, are shown by our old records to have been adopted and used in the exact form proposed, till considerations which had not been foreseen forced us to change to the form as printed. Even after years of

experience one is not safe in pronouncing on an apparent improvement without consulting the voluminous records of previous experiments. Even those who have used the system longest have sometimes been misled into adopting changes which on trial they were compeld to reject, going back to the original form at the cost and confusion of two changes.

The user who adopts the printed form avoids the criticism sure to be aimd at any possible scheme. The moment he makes one "improvement" he must defend all the heads or alter them to suit each critic. Much time is saved by saying that the scheme is used as printed, and blunders are the author's, not the user's. The cooperation of those interested is invited in eliminating any errors found either in tables or index.

The following brief notes show the most important variations found practicable in the "relativ index system," oftener called the "Decimal or Dewey system."

1 Letter notations for changes. To protect other users from confusion, the publishers insist, as the copyright entitles them to, that these numbers shall not be printed with changed meanings without some clear indication of the fact in the number itself. If the reasons that led us to adopt the form printed are not conclusiv to another, we wish to remove any obstacles to his use of the system with such changes as shall satisfy him. This can be readily done by using a letter, or some other character than the 10 digits, to mark changes, e. g., if you wish a different set of subdivisions under 630, make it out to suit, and number it 63a, 63b, 63c, etc. It will arrange in its exact place and exact order without difficulty, and no other user of the system will be confused by the numbers. In the index, cancel the 1, 2, 3, etc., you have discarded, and write in the a, b, c, etc., adopted. **Whenever you use our exact numbers, use also our meanings for them exactly as indext.** If you wish to change a head from one place to another, cancel it where it stands, and *leave that number blank* in the tables. Then insert the head in its new place by the use of a letter instead of the figure omitted as above, as if it had never been in the tables.

This plan of introducing letters or other symbols wherever each user pleases, will give all needed freedom to the personal equation and desire for originality, and meets all real wants for peculiar classification in peculiar cases.

Fiction and juvenils. In some cases it is usually best to modify class numbers by letters as above. In popular libraries half the circulation is often fiction. A great saving is effected here by omitting the class number entirely, as directed in these abridged tables, and using merely the book number, it being understood that *no* class number means fiction. After fiction, their great circulation makes juvenils a good place to economize, if they are kept separate. The

books are clast exactly as if for adults, then a J is prefixt to show their special character. This gives J alone as the class number for juvenil fiction, and J942, for a child's history of England. These are then arranged in a parallel library by themselves, so that J942 comes between J941, juvenil history of Scotland, and J943, juvenil history of Germany.

If at any time it seems desirable to abandon the separate J library, it can be done without altering a number, by distributing the J books among the regular classes, either ignoring the J entirely, or preferably by putting all the J books by themselves at the end of each class number.

There are thus three methods: 1, to have a separate J library; 2, to have the J books by themselves at the end of each class number; 3, to have the J books in alfabetic order among other books on the same subject. In this last case the J is useful only to call attention plainly to their juvenil character.

The same marking is used for all these plans, and one can be changed into the other by simply distributing the books the other way, and telling the attendants that it has been done. Juvenil books go with adult books of the same class, unless a special J library is made containing all juvenils.

Biografy. Similarly it is recommended that the large class of individual biografy, as to the best treatment of which there is great difference of opinion, should be markt B in place of the subject number. *See* p. 15, Book numbers; and tables "920 Biografy."

Instead of the plan printed in the tables, biografy may be distributed as far as possible among the subjects it illustrates, leaving, of course, under B lives not bearing specially on any subject; e. g. all lives of musicians go under 780 and its subdivisions, the life of Wagner being 782. Even in this arrangement it is better to indicate that it is biografy, putting B before the class number; i. e. mark Wagner's life B782. (See also Book numbers, p. 17.)

Parallel libraries. This treatment of fiction, juvenils, and biografy illustrates the principle of parallel libraries. Its other chief application is for reference and language collections. Some libraries have a constituency not reading English, and so need a parallel library in German, or Swedish, or French, etc. This is most easily made by simply prefixing the initial of the language to the class number. The German parallel library is

made by simply putting all G's together and arranging by class numbers. This plan has proved very satisfactory in actual use.

To separate books most needed for reference, mark R before class numbers, and arrange the books together, as an R library. When the books are to go into the general collection again, a line can be drawn thru this letter. In the same way it often happens that a general private library is given on condition that it be kept together; e. g, the Phœnix library of Columbia college. This has P prefixt to the class number, and thus is a parallel library by itself. An initial is better in such cases than a * or similar mark, since the initial helps the memory and is just as brief. The same plan applies, of course, if the library has an "inferno" for books not used without permits, or more distant rooms where books worth keeping but seldom calld for can be arranged in a parallel storage library.

Combining language and literature. The same principle can be applied also in combining each language with its literature, if it is preferd to abolish the class, Philology, and make it simply an appendix to Literature. The reverse would hold true if a philologist wisht to abolish Literature and make it an appendix to Philology; e. g. using 82p for English philology, and adding philology subdivisions, English dictionaries would become 82p3, English grammars 82p5, etc., and arrange either just before or just after English literature, 820, 821, etc. For a better plan see below, Broken order.

2 Broken order. Another common and often desirable variation for shelf arrangement is to break the sequence of the numbers, in order to get books most used nearest the delivery desk. The strict theory is to keep the whole 1000 numbers in exact sequence; but a higher rule to be obeyd everywhere is to sacrifice any theory for a substantial gain. Practically it will seldom happen that it is not best to break the order of the classes. Often divisions are best arranged out of numeric place; e. g. 520 Astronomy may be wanted in a room accessible at night; fiction, juvenils, and biografy are always wanted near the delivery desk in a public library, and in strict order are as likely to come at the most distant point. Various local reasons may make a broken order desirable. There need be no hesitation in adopting it if enough is gaind, but there should be charts clearly showing where each division starts; e. g. after 430 "Preceding 830;" after 520 "In observatory." The rule is to specify where books may be found that have been entirely removed from the general library arrange-

ment. The page of 100 divisions is reprinted on cards with wide margins for just this use. Opposit each division is markt its beginning on the shelves, and it is easy to vary the order as much as necessary, tho of course the nearer the divisions run in regular order 1 to 99, the easier it is for a stranger to find his way about. Wherever variations from the printed order are made, a wood or cardboard dummy in the regular place should have markt on its side the present location of the subject removed; e. g. "491 will be found preceding 891."

This broken-order plan is best for bringing together the philology and literature of each language without altering numbers or prefixing any letter. Let 420 be shelvd just ahead of 820, 430 ahead of 830, and so for all languages, making the general note that all 400's are shelvd just ahead of the corresponding 800's.

3 **Pro and con division of topics.** In many cases it is very useful to separate the books on a topic with strongly markt sides, so that either set of views and arguments may be seen by itself; e. g. 337, Protection and Free Trade. The number alone may be used for general works, giving facts, etc., and advocates and opponents may be separated by + and — for positiv and negativ, or by p and c, the initials for pro and con, words which, tho short, are too long for a circulating library to charge. In reference libraries, on cards, etc., most will prefer to write out *pro* and *con*, to distinguish the two groups. The order on the shelves is, of course, alfabetic, viz 337, $\overset{c}{337}$, $\overset{p}{337}$; or if + and — are used, the usual order is followd: +, —. These letters or signs should be written above or prefixt to the class number, not written 337c, 337—, etc., for the space at the end should be left free for other figures, if decided later to class more closely.

4 **Unassignd numbers.** The one and two figure numbers 1–99 are available for special uses without confusion with tables or index, for initial 0 is printed in the tables and index for all numbers before 100. In classification it sometimes happens that the first two figures are obvious at a glance, but time must be taken to determin the third. It is convenient to write these first figures, but if a mathematical book receives its first two figures (51), this unfinisht number is liable to be confused with the two-figure number 51. This danger may be largely avoided by writing the decimal point after a blank; e. g. 51 ., to show that a figure is omitted.

These four notes suggest the range of variations that may be made in the *Decimal classification* and illustrate its adaptability to widely different conditions. For book numbers which decide the order of material after it is groupt into its final classes. *see* page 17 and *Library notes* no. 11.

OTHER USES

Tho the system was devised for library catalog and shelf arrangement, 40 years' use has developt many new applications; but the essential character of the plan has remaind unchanged from the first. Nearly every administrativ department feels directly the great economy, and in every field of literary activity this classification has been found a labor-saving tool, whose practical usefulness has exceeded the most sanguin hopes of its early friends.

Bookstores. The plan is a great convenience to both dealers and customers, when applied to the miscellaneous stock of a bookstore. Very often a much-wanted book, specially if not recently publisht, is reported "not in stock," when such an arrangement by subjects would have reveald its place at once. Specialists often find on the shelves books that they would never have orderd, but are glad to buy after examination. Experience proves it a profitable thing for a dealer to arrange his books so that each person may find those in which he is interested without examining the entire stock.

Scrap books. The plan has proved of great service in preserving newspaper clippings. Scrap sheets of uniform size are used, with the class number of the subject written in the upper corner of the front edge of the page, and clippings are mounted on the sheet as in a common scrap-book. When one sheet is full, another is inserted at the exact place. Thus perfect classification is kept up without blank sheets, and at the smallest outlay of money and trouble. This "L. B. scrap book" is for most persons the best form. These sheets are then arranged numerically like the subject card catalog, the sheets of each class being farther arranged, when desirable, under alfabetic subheads. Scraps thus mounted are shelvd either in manila pamflet cases or in patent binders.

Index rerums. These are best made on the standard P size (7.5 x 12.5 cm), card or slip. Long study and experience prove that the stock L200[a] is best for private indexes, etc. It costs only half as much as the Br 400,[a] which is largely used for public library catalogs, takes only half the room, and gives great satisfaction.

Where durability and convenience of handling are less important than the strictest economy, common heavy writing paper may be used; but most novices greatly diminish the usefulness of the card system by using ordinary machine-cut cards or slips which vary in hight so much as to make quick and accurate manipulation impossible. The extreme variation that should be tolerated is 1 millimeter or $\frac{1}{25}$ in. This will be understood by placing a 7.4 cm card between two 7.5 cm cards. The fingers make a bridge across the taller cards and the lower one is missed entirely in rapid turning. Cards must be cut by accurate gages or they will lose

[a] This is a catalog number of the Library Bureau. L200 means linen ledger paper of which a square meter weighs 200 grams. Br 400 means bristol weighing 400 grams to the square meter.

half their value, and in many cases make it necessary to recopy the material at a cost tenfold greater than to have thrown away the imperfectly cut cards or slips.

The class number is written in the upper left corner, any alfabetic subject head follows at the right, and notes fill the card below. The cards are then filed in order of class numbers like the subject card catalog, the cards of each class being farther arranged, as in case of the scrap sheets, according to any alfabetic subheads. Scores of devices for convenient handling and storing of these slips and of scrap sheets and pamflets are already manufactured. Details will be found in the full descriptiv and illustrated catalog.

Another admirable form is similar to the L. B. scrap-book or shelf list; in fact, it is simply a shelf list without the printed head and colum lines. The subject number goes in the upper corner, and all notes on that subject are written on the sheet below. This book fits an ordinary shelf, has the advantage of a full letter page in sight at once, and holds over five times as much as the postal card. Of course the system can be applied to slips or sheets of any size, but there are literally hundreds of accessories and conveniences exactly adapted to these two sizes, which are used tenfold more than all others combined; so it is folly for one to begin on another size, and lose the advantages of this uniformity. If intermediate sizes must be had, the ones most used and least objectionable are Note, 12.5, x 20 cm, and Magazine, 17.5 x 25 cm. There are repeated cases where users of some other size have finally found it profitable to change to either the Postal, 7.5 x 12.5, or to the Letter, 20 x 25 cm, even at the cost of rewriting many notes.

Note books are best in this last described form. The much poorer method is to take a bound blank book, and assign the class numbers in order, giving about the space it is thought each will require, and, when the pages so assignd are full, note at the bottom of the page where the subject is continued. This has all the objections of the old fixt location as compared with the relativ, and will hardly be adopted by any person who has ever seen the simplicity and economy of the shelf-list system.

Topical indexes. The class numbers are used in indexing books read, usually by making the entries in the index rerum under proper numbers. The number takes the place of a series of words, and the results can be handled, arranged, and found much quicker because of the simple numerals.

The advantages of the system for making topical indexes of collected works, periodicals, transactions, etc. are evident. These consolidated indexes may be arranged together with the card catalog of the books, or by themselves, as seems best in each case.

It would exceed the limits of this brief description to notice all the varied applications of the system. Enough have been mentiond to show its wide adaptability to the wants of the librarian and the student.

FIRST SUMMARY

CLASSES

0 General works

1 Philosophy

2 Religion

3 Sociology

4 Philology

5 Natural science

6 Useful arts

7 Fine arts

8 Literature

9 History

SECOND SUMMARY

DIVISIONS

000 General works
010 Bibliography
020 Library economy
030 General cyclopedias
040 General collections
050 General periodicals
060 General societies
070 Newspapers
080 Special libraries. Polygraphy
090 Book rarities

100 Philosophy
110 Metaphysics
120 Special metaphysical topics
130 Mind and body
140 Philosophical systems
150 Mental faculties. Psychology
160 Logic. Dialectics
170 Ethics
180 Ancient philosophers
190 Modern philosophers

200 Religion
210 Natural theology
220 Bible
230 Doctrinal. Dogmatics. Theology
240 Devotional. Practical
250 Homiletic. Pastoral. Parochial
260 Church. Institutions. Work
270 Religious history
280 Christian churches and sects
290 Ethnic. Non-Christian

300 Sociology
310 Statistics
320 Political science
330 Political economy
340 Law
350 Administration
360 Associations and institutions
370 Education
380 Commerce. Communication
390 Customs. Costumes. Folklore

400 Philology
410 Comparative
420 English
430 German
440 French
450 Italian
460 Spanish
470 Latin
480 Greek
490 Minor languages

500 Natural science
510 Mathematics
520 Astronomy
530 Physics
540 Chemistry
550 Geology
560 Paleontology
570 Biology
580 Botany
590 Zoology

600 Useful arts
610 Medicine
620 Engineering
630 Agriculture
640 Domestic economy
650 Communication. Commerce
660 Chemical technology
670 Manufactures
680 Mechanic trades
690 Building

700 Fine arts
710 Landscape gardening
720 Architecture
730 Sculpture
740 Drawing. Decoration. Design
750 Painting
760 Engraving
770 Photography
780 Music
790 Amusements

800 Literature
810 American
820 English
830 German
840 French
850 Italian
860 Spanish
870 Latin
880 Greek
890 Minor languages

900 History
910 Geography and travels
920 Biography
930 Ancient history
Modern:
940 Europe
950 Asia
960 Africa
970 North America
980 South America
990 Oceanica and polar regions

General works

010 Bibliografy

011 General bibliografies. Universal catalogs

011 is by **Authors**. If by **Subjects** they go in 016.
012-016 include both bibliografies and catalogs. 017-019 is limited to catalogs of general collections.

012 Of individual authors

Alfabeted by bibliografees, not by compilers; e. g. Chaucer, Dante, Ruskin, etc.

013 Of special classes of authors

e. g. Books written by Jesuits, by Roman catholics, by members of the Bavarian academy, etc., or these may go with the church, society, etc. with references here.

014 Of special forms. Anonyms, pseudonyms, etc.

015 Of special countries

Books publisht in the country. Publishers' lists, current publications. Bibliografy of books publisht in England, as Lowndes or English catalog.
The history of literature, i. e. belles lettres, poetry, drama, fiction, etc. goes, of course, with those topics in 800; but the literary history of any given place or period covering the writings on all subjects as well as in literature, is bibliografy, and goes usually in 015, tho the literary history of some special class is 013.

016 Of special subjects

Bibliografy of

- .1 philosofy
- .2 religion
- .3 sociology
- .4 philology
- .5 natural science
- .6 useful arts
- .7 fine arts
- .8 literature
- .9 history

Library and sale catalogs

Catalogs of any special subject, whether subject, author, or dictionary, go under its subject number, in 016, which is the ruling heading wherever it conflicts with another. 017-019 therefore includes only catalogs of general collections, limited to no one class or subject

017 Clast catalogs. Systematic or logical

For all forms of alfabetic subject catalogs, see 019. See 016 for Bibliografies.

018 Author catalogs See 011 for Bibliografies.

A volume containing both author and subject catalogs is more useful in 017. 018 includes accession, chronologic, and any other forms (except subject and dictionary) of catalogs of collections.

019 Dictionary catalogs. Alfabetico-clast, etc.

020 Library economy

021 Scope and founding of libraries

022 Buildings See also 727, Architecture of educational buildings

023 Government and service

024 Regulations for readers

025 Administration. Departments

025 is for the librarian's part. The trustees build and furnish (022); make rules for government and service (023), and regulations for readers (024); but the administration involves questions of its own, which, however, are closely allied to topics in 021-024 and elsewhere; e. g. the librarian must know his side of binding, and be able to give proper directions and supervision, but need not know all the details of the binder's craft (686).

Administration includes Supervision; Acquisition; Utilization; Preservation; i. e. the librarian's duty to books is to Get, Use, Keep.

026 Libraries on special subjects

Histories, reports, statistics, bulletins, handbooks, circulars, and everything about the library not more required in one of the sections above; e. g. a Medical library, a Chess library, but the catalog of a Chess library is 016.7; blanks, etc. from any library go in 025, as more used in studying topics; but, if history of individual libraries is specialized, duplicates are also desirable under the library in 026-027, thus making a complete set of its publications.

027 General libraries

This includes both circulating and reference; i. e. all not limited to special subjects.

028 Reading and aids

See also 374, Self education.

029 Literary methods and labor-savers

Much in 025 and 028 belongs equally under 029 in its full meaning, but practical convenience is best servd by keeping such material under those heads instead of separating a part here.

030 General cyclopedias

031 American

e. g. Appleton, International, Johnson.

032 English

e. g Britannica, Chambers.

033 German

e. g. Brockhaus

034 French

e. g. Larousse.

035 Italian

036 Spanish

037 Slavic

038 Scandinavian

039 Minor languages

040 General collected essays

041 American
042 English
043 German
044 French
045 Italian
046 Spanish
047 Slavic
048 Scandinavian
049 Minor languages

050 General periodicals, magazines

051 American
e. g. Atlantic, Century, Harper's, Scribner's.
052 English
e. g. Athenæum, Blackwood's, Contemporary.
053 German
054 French
055 Italian
056 Spanish
057 Slavic
058 Scandinavian
059 Minor languages

060 General societies, transactions

061 American
062 English
063 German
e. g. Academies of Berlin, Vienna, etc.
064 French
065 Italian
066 Spanish
067 Slavic
068 Scandinavian
069 Minor languages

070 General newspapers. Journalism

071 American

e. g. Albany argus, N. Y. tribune, Nation, Harper's weekly.

072 English

e. g. Times (London), Saturday review, Spectator.

073 German

074 French

075 Italian

076 Spanish

077 Slavic

078 Scandinavian

079 Minor languages

080 Special libraries. Polygrafy

081

082

083

084

085

086

087

088

089

Left blank to be used (if preferd to prefixing an initial) for general collections of books which by terms of gift or for other cause must be kept together. This must be kept distinct from 040 where the individual books are polygrafic, i. e. are bound pamflets, essays, addresses, scrap or note books, etc., too general in scope to go under any single class. Juvenils could be put here, but are better kept in a separate class, markt J.

090 Book rarities

Books about these topics, and those chiefly valuable because of their rarity, go here. A rare early edition of Shakspere goes in 822 with reference from 094. 090 is mostly used for grouping references to books located elsewhere.

091 Manuscripts. Autografs

Photografs of mss go with subject rather than here.
For Diplomatics and Paleografy, see 417, and under the special language 421, 431, etc.

092 Block books

093 Early printed books. Incunabula

094 Rare printing

Aldines, Elzevirs, Caxtons, etc.; privately printed books: unique books.

095 Rare binding

Noted binders, costly ornament, curious bindings.

096 Rare illustrations or materials

Illuminated. Illustrated by inserted plates. Printed on vellum, silk, bark, etc., in gold or silver letters, etc.

097 Ownership. Book plates. Ex libris

098 Prohibited. Lost. Imaginary

099 Other rarities. Curiosa. Minute size, etc.

Philosofy

100 Philosofy, general works

Works limited to none of the nine divisions.

101 Utility
102 Compends. Outlines
103 Dictionaries. Cyclopedias
104 Essays. Lectures. Addresses
105 Periodicals. Magazines. Reviews
106 Societies. Transactions. Reports
107 Education. Study and teaching
108 Polygrafy. Collected works. Extracts, maxims, etc.
109 History of philosofy

110 Metaphysics

111 Ontology
Nature of being. Substance and form.
112 Methodology
Philosofical classification of knowledge. Terminology. For book classification see 025.4.
113 Cosmology
114 Space
115 Time
116 Motion
117 Matter
118 Force
119 Quantity. Number

120 Other metaphysical topics

121 Theory of knowledge. Origin. Limits
122 Causation. Cause and effect
123 Liberty and necessity See also 159, Will.
124 Teleology. Final causes
125 Infinit and finite
126 Consciousness. Personality
127 Unconsciousness. Automata
128 The soul See also 218, Natural theology; 237, Future state.
129 Origin of the individual soul

130 Mind and body. Anthropology

131 Mental physiology and hygiene
132 Mental derangements
133 Delusions. Witchcraft. Magic
134 Mesmerism. Animal magnetism. Clairvoyance
135 Sleep. Dreams. Somnambulism
136 Mental characteristics See also 131, Mental physiology.
137 Temperaments
138 Physiognomy
139 Phrenology. Mental photografs, etc.

140 Philosofic systems

General discussion of each system; do not attempt to label each writer as an exponent of some system. Collected philosofic works of individual authors are clast in the sections, 180-199; class individual work according to subject.

141 Idealism. Transcendentalism
142 Critical philosofy
143 Intuitionalism See also 156, Intuitiv faculty; 171, Ethical theories.
144 Empiricism
145 Sensationalism
146 Materialism. Positivism
147 Pantheism. Monism See also 212, Natural theology.
148 Eclecticism
149 Other philosofic systems
See also 211, Rationalism, Scepticism; 214, Fatalism.

150 Mental faculties. Psychology

For mind in animals, see 591.

151 Intellect. Capacity for knowing
152 Sense. Sense perceptions Passiv or receptiv faculty.
153 Understanding Activ or thinking faculty.
154 Memory. Reproductiv power
155 Imagination. Creativ power
156 Intuitiv faculty. Reason
Reasoning, the act of deriving conclusions from premises, 153.
157 Sensibility. Emotions. Affections
158 Instincts. Appetites
159 Will

160 Logic. Dialectics

See also 153, Reasoning power. For Logic of chance, see 519, Probabilities.

161 Inductiv

162 Deductiv

163 Assent. Faith See also 234, Doctrin of salvation.

164 Symbolic. Algebraic Logical machines.

Logical topics

165 Sources of error. Fallacies

166 Syllogism. Enthymeme

167 Hypotheses

168 Argument and persuasion

169 Analogy. Correspondence See also 219, Natural theology.

170 Ethics, theoretic and applied

Many topics in applied ethics occur also in law, specially in 343, Criminal law. See also 377, Ethical education.

171 Theories. Philosofy of ethics

172 State ethics

173 Family ethics

174 Professional and business ethics

175 Ethics of amusements For Amusements, see 790.

176 Sexual ethics

177 Social ethics. Caste

178 Temperance. Stimulants and narcotics

See also 613, Hygiene; 331, Laboring classes.

179 Other ethical topics.

180 Ancient philosofers

Sections 180-199 are for the history of philosofy in special countries and for the discussion of the philosofic systems of individual authors and for their works not clearly belonging elsewhere. Mill's *Logic* is 160 not 192, and Hegel's *Philosophy of history* is 901 not 193, but their complete philosofic works belong here. Collected lives of philosofers are clast 921, and individual lives are placed in the alfabetic series of biografy.

181 Oriental philosofers

182 Early Greek philosofers

183 Sophistic and Socratic philosofers

184 Platonic. Older Academy

185 Aristotelian. Peripatetic. Lyceum

186 Pyrrhonist. New Platonist

187 Epicurean. Epicurus. Lucretius

188 Stoic

189 Early Christian and medieval philosofers

190 Modern philosofers

See notes under 140 and 180.

191 American philosofers

192 British philosofers

193 German philosofers

194 French philosofers

195 Italian philosofers

196 Spanish philosofers

197 Slavic philosofers

198 Scandinavian philosofers

199 Other modern philosofers

Religion

200 Religion. General works

201 Philosofy. Theories

202 Compends. Outlines

203 Dictionaries. Cyclopedias

204 Essays. Lectures. Addresses See 252, Sermons.

205 Periodicals. Magazines. Reviews

206 Societies. Transactions. Reports

Bible, Tract, etc. societies; history, reports, etc., but their publications go of course with subject.

207 Education. Theological seminaries. Training schools See also 377, Religious and secular education.

208 Polygrafy. Collected works. Extracts, maxims, etc.

Many collections go under a more specific head; e. g. 240.

209 History of religion

210 Natural theology

211 Deism and atheism

Scepticism. Infidelity. Rationalism, etc.
See 239, Apologetics; 231, God; 273, Heresy.

212 Pantheism. Theosofy See 147, Pantheistic philosofy: 129, Origin of soul.

213 Creation. See 113, Cosmology. Evolution See 575, Science.

214 Providence. See 231, Christian doctrin of God. Theodicy. Fatalism See 231, for Christian view.

215 Religion and science See 239, Apologetics.

216 Evil. Depravity See 149, Pessimism; 233, Doctrin of man.

217 Prayer. Prayer gage See 264, Public worship; 248, Personal religion.

218 Future life. Immortality. Eternity

See 237, Future state; 128, The soul.

219 Analogies. Correspondences See also 169, Logic.

220 Bible. General works

For similar works limited to Old or New Testament, or individual books, see specific head below.

.1 **Canon. Inspiration. Profecy**

.2 **Concordances. Analyses**

.3 **Dictionaries. Cyclopedias**

.4 **Original texts and early versions. Codices**

.5 **Versions of Bible. Polyglots**

These are translations from the original Hebrew and Greek. Translations from other early texts, e. g. an English translation of the Syriac, go in 220.4.

220.6 **Hermeneutics. Exegesis. Symbolism. Typology**

.7 **Commentaries on whole Bible, and annotated editions**

For notes, etc., on parts of the Bible, see the most specific head in 221-229.

.8 **Special topics** e. g. Biblical astronomy, botany, etc.

.9 **Biblical geografy and history**

221 Old Testament. Texts. Introductions, etc.

222 Historical books. Genesis to Esther

223 Poetic books. Job to Song of Solomon

224 Profetic books. Isaiah to Malachi

225 New Testament. Texts, introductions, etc.

226 Gospels and Acts

227 Epistles

228 Apocalypse

229 Apocryfa

230 Doctrinal. Dogmatics. Theology

231 God. Unity. Trinity

232 Christology

233 Man

234 Salvation. Soteriology

235 Angels. Devils. Satan

236 Eschatology. Last things

237 Future state See 218, Natural theology.

238 Creeds. Confessions. Covenants. Catechisms

239 Apologetics. Evidences of Christianity

240 Devotional. Practical

241 Didactic For Catechisms, see 238.

242 Meditativ. Contemplativ See 248, Personal religion.

243 Hortatory. Evangelistic

244 Miscellany. Religious novels, Sunday-school books, allegories, etc.

But class Bunyan in fiction, because of his literary prominence.

245 Hymnology. Religious poetry

See 223, Psalms; 264, Public worship.

246 Ecclesiology. Symbolism. Religious art

246-247 cover religious bearings. For art side see 700.

247 Sacred furniture. Vestments. Vessels. Ornaments, etc.

248 Personal religion. Asceticism See also 273, Heresies.

249 Family devotions See 264, Public worship.

250 Homiletic. Pastoral. Parochial

251 Homiletics. Preaching See 264, Public worship.

252 Sermons

Sermons on specific topics are more useful, like other pamflets, clast with the topics, e. g. a sermon on family devotions in 249, not 252; on strikes in 331.

253 Pastoral visitations, evangelistic

254 Clerical support. Celibacy

See also 176, Sexual ethics; 248, Asceticism.

255 Brotherhoods. Sisterhoods

In the parish. For Monastic orders, see 271.

256 Societies for parish work. Gilds. Sodalities

Local societies. Discussion of desirability of such work. For general societies, see 206 and 267.

257 Parochial schools. Libraries, etc.

See 377, Religious and secular education; 027, Libraries.

258 Parish care of sick, fallen, etc.

See also 176, Sexual ethics.

259 Other ministrations and work

260 Church. Institutions and work

261 The church Its influence on morals, civilization, etc.

262 Ecclesiastical polity

263 Sabbath. Lord's day. Sunday

264 Public worship. Divine service. Ritual

See also 246, Ecclesiology; 247, Sacred furniture.

265 Sacraments. Ordinances

266 Missions. Home and foreign

Missions in special countries or places, including those of all churches, go under the geografically divided Religious history, 274-279, with references under 266.

267 Associations. Y. M. C. A., etc.

268 Sunday schools

269 Revivals. Retreats. Parish missions

270 Religious history

271 Monastic orders

See also 255, Parish brotherhoods; 377, Religious and secular education.

272 Persecutions

See also special sects, 280; and history of special countries, 940-999.

273 Heresies For the history of special doctrins, see 230-239, Doctrinal theology

274 General religious history of Europe

General religious history of

.1 **Scotland. Ireland**

.2 **England. Wales**

.3 **Germany. Austria**

.4 **France**

.5 **Italy**

.6 **Spain. Portugal**

.7 **Russia**

.8 **Norway, Sweden and Denmark**

.9 **Minor European countries**

275 General religious history of Asia

General religious history of

.1 China
.2 Japan
.3 Arabia
.4 India
.5 Persia
.6 Turkey in Asia
.7 Siberia
.8 Afghanistan. Turkestan. Baluchistan
.9 Farther India

276 General religious history of Africa

General religious history of

.1 North Africa
.2 Egypt
.3 Abyssinia
.4 Morocco
.5 Algeria
.6 North central Africa
.7 South central Africa
.8 South Africa
.9 Madagascar. Mauritius

277 General religious history of North America

General religious history of

.1 Canada. British America
.2 Mexico. Central America
.3 United States
.4 Northeastern or North Atlantic states, New England
.5 Southeastern or South Atlantic states
.6 South central or Gulf states
.7 North central or Lake states
.8 Western or Mountain states
.9 Pacific states

For the exact states covered by these general designations, see 974-979.

278 General religious history of South America

General religious history of

.1 Brazil
.2 Argentine Republic. Patagonia
.3 Chili
.4 Bolivia
.5 Peru
.6 U. S. of Colombia. New Granada. Ecuador
.7 Venezuela
.8 Guiana
.9 Paraguay. Uraguay

279 General religious history of Oceanica. Polar regions

General religious history of

- .1 **Malaysia**
- .2 **Sunda**
- .3 **Australasia**
- .4 **Australia**
- .5 **New Guinea**
- .6 **Polynesia**
- .7 **Isolated islands**
- .8 **Arctic regions**
- .9 **Antarctic regions**

280 Christian churches and sects

See also 270, Religious history.

281 Primitiv and oriental churches

282 Western or Roman catholic church

283 Anglican and American P. E. church

284 Continental protestant sects. Protestantism

285 Presbyterian. Reformd. Congregational

286 Baptist

287 Methodist

288 Unitarian See also 273, Heresies.

289 Minor Christian sects

290 Ethnic. Non-Christian

291 Comparativ and general mythology

292 Greek and Roman mythology

293 Teutonic and northern mythology

294 Brahmanism. Buddhism For other Indic religions, see 299.
See also 891, Sanskrit literature; 177, Social ethics, Caste.

295 Parseeism. Zend Avesta For other Iranic religions, see 299.
See also 891, Zend literature.

296 Judaism For other Semitic religions, see 299.

297 Mohammedanism

298 Mormonism See also 173, Family ethics.

299 Minor non-Christian religions

Sociology

300 Sociology. General works

301-309 all have Sociology in general as their subject, but it is treated in these various forms. A periodical on education goes in 370, not 305, which is only for periodicals on sociology in general. In sociology, most works in these forms are limited to one division; e. g. to political economy, education, law, etc.

301 Philosofy Theories

See 901, Philosofy of history.

302 Compends. Outlines

303 Dictionaries. Cyclopedias

304 Essays. Lectures. Addresses

305 Periodicals. Magazines. Reviews

306 Societies. Transactions. Reports

307 Education. Study and teaching See also 370.

308 Polygrafy. Collected works. Extracts, etc.

Put here collected works of statesmen; e. g. works of Adams, Jefferson, etc.

309 History of social science

310 Statistics

311 Theory, methods. Science of statistics

312 Population. Progress of. Vital statistics. Births. Deaths. Mortality. Longevity

See also 614, Public health.

313 Special topics

The statistics of any special matter are put with the subject, e. g. of Domestic animals in 636, of Shorthand in 653, of French novels in 843, of Theaters in 792; etc. The statistics of New York city would be 317, but the statistics of Medicin in New York would be put with 610, Medicin, i. e. the topic outranks the locality.

Statistics of

314 Europe

315 Asia

316 Africa

317 North America

318 South America

319 Oceanica

320 Political science

321 Form of state
This is for discussion of form of state, what should be; for what is, see 342, Constitutional law and 353-354, Administration.

322 Church and state See also 172, State ethics; 261, The church.

323 Internal or domestic relations
Free speech, Liberty of the press; for Morals of the press, see 179; also, 070, Newspapers.

324 Suffrage

325 Colonies and immigration

326 Slavery See also 973.7, Civil war.

327 Foreign relations

328 Legislativ bodies and annals

329 Political parties. Party conventions

330 Political economy

331 Capital. Labor and wages

332 Banks. Money. Credit. Interest

333 Land. Ownership, rights and rent
See also 338, Production; and 630, Agriculture.

334 Cooperation

335 Socialism and communism

336 Finance

337 Protection and free trade

338 Production. Manufacture. Prices

339 Pauperism. Consumption of wealth
See also 362, Asylums; 331, Labor and wages.

340 Law Most periodicals belong in Private law, 347.

Public law

341 International law

342 Constitutional law and history For Administrativ law, see 350

343 Criminal law

344 Martial law

Private law

345 U. S. statutes and cases

346 British statutes and cases
Includes all reports in English language except U. S. reports.

347 Treatises. American and English private law
Put law of special topics with the subject; e. g. Insurance law, 368.

348 Canon law

349 Foreign law

350 Administration Including Military science.

351 Administration of central government

352 Local government. Town. City. County
See also 379, Public schools; 020, Libraries.

353 United States and state government
See also 342, Constitutional law and history.

354 Organization of central government. Foreign states
See also 342, Constitutional law and history.

355 Army. Military science See also 623, Military engineering.
Limited to Military science, tactics, etc. For War department of U. S. or history of regiments, see 353.

356 Infantry

357 Cavalry

358 Artillery, engineers, etc.

359 Naval science See 623, Naval engineering.

360 Associations and institutions
This covers what is best known as "Charities and corrections," and general works on this topic go in 360.

361 Charitable

362 Hospitals, asylums, and allied societies
For inebriate asylums, see 178, Temperance.

363 Political Tammany, Primrose leag, Ku Klux, etc.

364 Reformatory Schools. Discharged convicts. Criminal classes.

365 Prisons. Disciplin

366 Secret societies See also 371, College secret societies.

367 Social clubs

368 Insurance See also 334, Cooperation.

369 Other Cincinnati.

370 Education

371 Teachers, methods, and disciplin

372 Elementary education. Primary schools

373 Secondary. Academic. Preparatory
The discussions of theories, methods, etc. are clast regardless of the source of support but schools clast in 372, 373, 376, 377 and 378 are all private or endowd except parts of primary, schools for women, and colleges, which, even if supported by public funds, are preferably clast in 372, 376 and 378. All other "public" schools, i. e. paid for from the public treasury, are groupt under 379. Besides schools such topics as "American students in Europe" go naturally in 373 or, if the topic be "European universities," in 378. Peabody and Slater funds go in 373

374 Self-education and culture. Extension teaching

375 Curriculum

376 Education of women See also 396, Woman's position and treatment.

377 Religious, ethical, and secular education

378 Colleges and universities With power of conferring degrees.

379 Public schools. State education

380 Commerce, communication

The technical side of these questions goes mostly in 650, Useful arts. Here belong discussions of social and political relations; e. g. government control of railroads, telegrafs, etc

381 Domestic trade

382 Foreign trade. Consular reports

383 Post-office See also 353, U. S. administration.

384 Telegraf. Cable. Telefone
See also 654, Telegrafic instruments and construction

385 Railroad and express See also 656, Transportation.

386 Canal and highway transportation ,,

387 River and ocean transportation ,,

388 City transit See also 625, Railroad and road engineering.

389 Weights and measures. Metrology See also 658, Business manuals and methods.

390 Customs. Costumes. Popular life

These heads are for discussions by topics. Customs, etc. of any special country go in 913-919. Books on a special topic in a special country go here, as the grouping by topics is the more important; e. g. Marriage in Japan is 392, but Japanese customs, 915.2. For Customs of primitiv man, see 571.

391 Costume and care of person See 646, Clothing; 613, Hygiene

392 Birth, home, and sex customs

393 Treatment of the dead See 614, Public health.

394 Public and social customs

395 Etiquet

396 Woman's position and treatment
For Costumes, see 391; Biografy, 920.7.
If a special library about women is wisht, 396 is the best place for it. Suffrage, education, and employment can then be put here with references from 324, 376, 331, etc.; but it would be unwise to bring everything about women here, e. g. to remove 618. Diseases of women, from the rest of medicin. Books on woman in general go in 396.

397 Gipsies. Nomads. Outcast races
People without nationalities who do not coalesce with the ruling people among whom they live. This includes Gipsy language, which has no place in the linguistic groups of 400, as the Gipsy people have no place in the geografic divisions of history.

398 Folk-lore. Proverbs, etc. See also 291-293, Mythology.
This section is for material needed in studying folk-lore. Mere stories for children, unless having a value to students of folk-lore, go in fiction, or in J, if there be a juvenil collection.

399 Customs of war
Weapons. War dances. Treatment of captivs. Scalping. Mutilation. Burning. Cannibalism.

Philology

400 Philology. General works

In Philology the general works put under 400-419 deal almost entirely with the Indo European languages. They are put here because they cover most of the divisions of this class, and in practice are most convenient here. Under 439, 479, and 489 are placed books limited to the Teutonic, Romance or Hellenic groups, and under 491 are placed only such general works as are specifically limited to the Indo-European group.

401 Philosofy. Origin of language
402 Compends. Outlines
403 Dictionaries. Cyclopedias
404 Essays. Lectures. Addresses
405 Periodicals. Magazines. Reviews
406 Societies. Transactions. Reports
407 Education. Schools of languages
408 Polygrafy. Collected works. Extracts, etc.
409 History of language

410 Comparativ

410-419 includes *comparativ* works in *general* and also those on the Indo-European group in general, but general *and comparativ* works on Teutonic group are 439; on Romance group 479; on Hellenic group 489. Everything about an individual language is put with that language.

411 Orthografy. Orthoepy. Alfabets
412 Etymology. Derivation
413 Dictionaries. Lexicografy
414 Phonology. Visible speech
415 Grammar. Morfology. Syntax
416 Prosody
417 Inscriptions. Paleografy

Rare early mss are put in 091; inscriptions of an individual language are clast with its orthografy; Latin inscriptions, 471; Greek inscriptions, 481, etc.

418 Texts
419 Hieroglyfics See also 493, Egyptian hieroglyfics.

420 English philology

421 Orthografy

See also this head treated in general grammars placed under 425.

422 Etymology. Derivation

422 is limited to derivation. For inflection, also called etymology, see 425.

423 Dictionaries. Idioms

Put a dictionary of two languages with the less known language. Under 423 put only English-English dictionaries. Put an English-French dictionary with French, 443; a French-Latin dictionary with Latin, 473. If in several languages, put with 413, or with least known. This plan brings together under each of the less known, all the dictionaries for translating either into or from that language. Some prefer to put each dictionary under the first language; i. e. that by which it is alfabeted. This gives under *each* language, regardless of its familiarity, *all* dictionaries for translating *from* it, but none for translating *into* it. These must be sought under the language from which the translation is to be made. For a cosmopolitan library this plan is simplest and best; but in an English library, the first plan, with only English dictionaries in 423, and both *in* and *out* dictionaries together under little known tongues is more convenient. References in either case show what may be found in the other place.

424 Synonyms. Homonyms

425 Grammar

425 includes general works, covering also orthografy and prosody.

426 Prosody

See also the head prosody, in general grammars, 425.

427 Dialects. Early forms. Slang

428 School books. Texts for learning the language

Including only books for *learning* the language, with grammatic or philologic notes, etc. For other works see the literature of the language, 820.

429 **Anglo-Saxon**

430 German

431 Orthografy

432 Etymology. Derivation

433 Dictionaries. Idioms

434 Synonyms. Homonyms

435 Grammar

436 Prosody

437 Dialects. Early forms. Slang

438 School books. Texts for learning the language

439 **Minor Teutonic** Including general works on Teutonic group.

440 French

441 Orthografy

442 Etymology. Derivation

443 Dictionaries. Idioms

444 Synonyms. Homonyms

445 Grammar

(French)

446 Prosody
447 Dialects. Early forms. Slang
448 School books. Texts for learning the language
449 **Provençal** Old French, as the earliest form of the language, is 447.

450 Italian

451 Orthografy
452 Etymology. Derivation
453 Dictionaries. Idioms
454 Synonyms. Homonyms
455 Grammar
456 Prosody
457 Dialects. Early forms. Slang
458 School books. Texts for learning the language
459 **Wallachian**

460 Spanish

461 Orthografy
462 Etymology. Derivation
463 Dictionaries. Idioms
464 Synonyms. Homonyms
465 Grammar
466 Prosody
467 Dialects. Early forms. Slang
468 School books. Texts for learning the language
469 **Portuguese**

470 Latin

471 Orthografy
472 Etymology. Derivation
473 Dictionaries. Idioms
474 Synonyms. Homonyms
475 Grammar
476 Prosody
477 Dialects. Early forms. Slang
478 School books. Texts for learning the language
479 **Minor Italic. Medieval Latin**
Including general works on Romance group.

480 Greek

481 Orthografy

482 Etymology. Derivation

483 Dictionaries. Idioms

484 Synonyms. Homonyms

485 Grammar

486 Prosody

487 Dialects. Early forms. Slang

488 School books. Texts for learning the language

489 **Minor Hellenic. Modern Greek**

Including general works on Hellenic group.

Works on Latin and Greek together, unless very clearly most useful with Latin, go with Greek, which here as elsewhere is made to include general works on the ancient classics.

490 Minor languages

491 Minor Indo-European

(Besides Teutonic, 420-439; Italic, 440-479; and Hellenic, 480-489.) This head, 491, includes general works on the Indo-European tongues, but general works on the Teutonic languages go in 439, on the Romance group in 479, on the Hellenic group in 489, while most of the material placed under 400-419 is really Indo-European; but see also 410.

492 Semitic

493 Hamitic

See 419, Hieroglyfics.

494 Scythian. Ural-Altaic. Turanian

Dravidian. Tamil. Finnish. Turkish.

495 Eastern Asiatic. Chinese

496 African

Excluding 493 Hamitic, 492 Semitic, etc. included in families above.

497 North American

498 South American

499 Malay-Polynesian and other

The Gipsy language is not assignd to any group, but is placed for convenience with 397, Gipsies.

Natural science

500 Science. General works

501-509 all have *Science in general* as their subject, but it is treated in these various *forms*. A periodical on *chemistry* goes in 540, not in 505, which is only for periodicals on science in general.

501 Philosofy. Theories. Utility

502 Compends. Outlines. Ancient and medieval science

503 Dictionaries. Cyclopedias

504 Essays. Lectures. Addresses

505 Periodicals. Magazines. Reviews

506 Societies. Transactions. Reports

507 Education. Teaching and studying. Museums

See also 370.

508 Polygrafy. Collected works. Extracts, etc.

.3 General scientific travels and surveys

Scientific travel in

.4 Europe

.5 Asia

.6 Africa

.7 North America

.8 South America

.9 Oceanica

509 History of science

510 Mathematics

511 Arithmetic See also 372, Elementary education

512 Algebra

513 Geometry

Descriptiv geometry is 515. See also Mathematical drawing, 744.

514 Trigonometry

515 Descriptiv geometry and projections

See also 744, Mathematical drawing

516 Analytic geometry

517 Calculus

519 Probabilities

520 Astronomy

521 Theoretical astronomy

Mathematical investigation of celestial motions, specially of the solar system. The motions of individual bodies are clast in 523.

522 Practical and sferical

523 Descriptiv astronomy

524 Maps, observations and tables

Series of observatory publications may be kept together under 524, or under 522 with the history and reports of the observatory. Special maps or observations; e. g. on sun or moon are better placed under 523. This number provides a place in case it is wisht to keep *all* maps and observations together

525 Earth

526 Geodesy and surveying

527 Navigation

528 Efemerides

529 Chronology

530 Physics

531 Mechanics

532 Liquids. Hydrostatics. Hydraulics

533 Gases. Pneumatics

534 Sound. Acoustics

535 Light. Optics

536 Heat

537 Electricity

538 Magnetism

For "Animal magnetism," Mesmerism, etc. see 134.

539 Molecular physics

540 Chemistry

541 Theoretic chemistry

542 Practical and experimental chemistry

543 Analysis

Of special materials. See also 614.3, Adulterations.

544 Qualitativ analysis

545 Quantitativ analysis

546 Inorganic chemistry

547 Organic chemistry

548 Crystallografy

549 Mineralogy

550 Geology

551[1] Physical and dynamic geology

Including physical geografy. For cosmic geografy, see 523.

552[2] Lithology. Petrografy. Petrology

553 Economic geology See also Mining engineering.

While the history of all the other products goes here, the history of the metals is more useful in 669, Metallurgy.

554 Geology of Europe

(Geology of)

.1 Scotland. Ireland
.2 England. Wales
.3 Germany. Austria
.4 France
.5 Italy
.6 Spain. Portugal
.7 Russia
.8 Norway, Sweden and Denmark
.9 Minor European countries

555 Geology of Asia

(Geology of)

.1 China
.2 Japan
.3 Arabia
.4 India
.5 Persia
.6 Turkey in Asia
.7 Siberia
.8 Afghanistan. Turkestan. Baluchistan
.9 Farther India

556 Geology of Africa

(Geology of)

.1 North Africa
.2 Egypt
.3 Abyssinia
.4 Morocco
.5 Algeria
.6 North central Africa
.7 South "
.8 " Africa
.9 Madagascar. Mauritius

557 Geology of North America
.1 Canada
.2 Mexico. Central America West Indies
.3 United States
.4 Northeastern or North Atlantic states. New England
.5 Southeastern or South Atlantic states
.6 South central or Gulf states
.7 North central or Lake states
.8 Western or Mountain states
.9 Pacific states

(Geology of)

For the exact states covered by these general designations, see 974-979.

558 Geology of South America
.1 Brazil
.2 Argentine Republic. Patagonia
.3 Chili
.4 Bolivia
.5 Peru
.6 U. S. of Colombia. New Granada Ecuador
.7 Venezuela
.8 Guiana
.9 Paraguay. Uruguay

(Geology of)

559 Geology of Oceanica. Polar regions
.1 Malaysia
.2 Sunda
.3 Australasia
.4 Australia
.5 New Guinea
.6 Polynesia
.7 Isolated islands
.8 Arctic regions
.9 Antarctic regions

(Geology of)

560 Paleontology

561 Plants
562 Invertebrates
563 Protozoa. Radiates
564 Mollusks
565 Articulates
566 Vertebrates
567 Fishes. Batrachia
568 Reptils. Birds
569 Mammals

570 Biology

571 Prehistoric archeology

For customs, see 390. For archeology of special countries, see 913.

572 Ethnology. Anthropology

See also 136, Mental race characteristics.

573 Natural history of man

574 Homologies

575 Evolution See also 213, Creation; 215, Religion and Science; 239, Apologetics.

576 Origin and beginnings of life

577 Properties of living matter

578 Microscopy See also 535, Optics.

579 Collectors manuals

580 Botany

581 Physiologic and structural botany

.9 Geografic

This is to be used only for general works and cross references. The "Flora of North America" is put 581.9; but "North American cryptogams" is clast 586 with a reference from 581.9, but No. Am. phanerogamia goes with 581.9 because it so nearly covers the subject.

General works covering both Phanerogamia and Cryptogamia are put under 580, as books on Vertebrates and Invertebrates are put under 590.

582 Phanerogamia See also Trees 715.

583 Dicotyledonae

584 Monocotyledonae

585 Gymnospermae

586 Cryptogamia

587 Pteridofyta

588 Bryofyta

589 Thallofyta

590 Zoology

591 Physiologic zoology

592 Invertebrates

593 Protozoans. Radiates

594 Mollusks

595 Articulates

596 Vertebrates

597 Fishes. Pisces

598 Birds. Reptils

.1 **Reptils**

599 Mammals. Mammalia

Useful arts

600 Useful arts. General works

601 Philosofy. Theories, etc.

602 Compends. Outlines

603 Dictionaries. Cyclopedias

604 Essays. Lectures. Addresses

605 Periodicals. Magazines. Reviews

606 Societies. Transactions. Reports. Fairs. Exhibitions

Special exhibitions go with their topics. This is general only.

607 Education. Schools of technology

608 Patents. Inventions

609 History of useful arts

For its history, see each special department.

610 Medicin

611 Anatomy. Histology

For Malformations see 617, Surgery. See also 591, Physiologic zoology.

612 Physiology

For vital phenomena in general, see 577, Properties of living matter. See also 591, Physiology of animals.

613 Personal hygiene

Care of health; Prophylaxis; Individual health; Laws of health.

.7 **Gymnastics and athletics. Recreation and sleep**

614 Public health

Public hygiene. Public sanitation. State medicin. Preventiv medicin. See also 628, Sanitary engineering; 352, Local administration.

.3 **Adulterations. Inspection of articles liable to affect public health.** Public analysts. State laboratories. For chemical analysis see 543.

.8 **Protection of human life from accidents, casualties, etc.**

615 Materia medica and therapeutics

616 Pathology. Diseases. Treatment

617 Surgery

618 Diseases of women and children. Obstetrics

619 Comparativ medicin. Veterinary

620 Engineering

621 Mechanical engineering

.1 **Steam engineering**

.3 **Electric engineering**

.4 **Air and gas engins**

Internal combustion engins.

622 Mining engineering

623 Military and naval engineering. Shipbuilding

For Military and naval science, Maneuvers, Strategy and tactics, see 355 to 359

624 Bridges and roofs

625 Railroad and road engineering

See also 624, Bridges.

626 Canal engineering

627 River, harbor and general hydraulic engineering

628 Sanitary engineering. Waterworks

629 Other branches of engineering

630 Agriculture

631 Soil. Fertilizers. Drainage

632 Pests. Hindrances. Blights. Insects

633 Grains. Grasses. Fibers. Tea. Tobacco, etc.

634 Fruits. Orchards. Vineyards. Forestry

See also 715, Landscape gardening.

635 Kitchen garden See 716, for Flower gardens.

636 Domestic animals See also 619, Veterinary medicin; 599, Zoology.

637 Dairy. Milk. Butter. Cheese

638 Bees. Silkworms

639 Fishing. Trapping

640 Domestic economy.

641 Food. Cookery. Gastronomy

See also 613, Hygiene ; 663, Beverages.

642 Serving. Table. Entertaining

See also 394, Customs.

643 Shelter: house, home

644 Heat. Light. Ventilation

See also 697, Heating ; 665, Gas, candles, lamps ; 621.3, Electric engineering.

645 Furniture. Carpets. Upholstery. Decoration

646 Clothing. Toilet. Cosmetics

See also 613 ; Hygiene ; 391, Costume and care of person.

647 Household organization and administration

Expenses. Servants: training, duties, wages. Hotels, clubs, flats, etc.

648 Sanitary precautions: laundry, cleaning

649 Nursery. Children. Sickroom

650 Communication. Commerce

651 Office equipment and methods

652 Writing. Materials, typewriters, cifer

653 Abbreviations. Shorthand

654 Telegraf. Cables. Signals For codes see 652.

655 Printing. Publishing. Copyright

656 Transportation. Railroading, etc.

657 Bookkeeping. Accounts

658 Business manuals. Methods. Tables

659 Advertising and other topics

660 Chemical technology

661 Chemicals

662 Pyrotechnics. Explosivs. Matches, etc.

663 Beverages See also 178, Temperance; 614.3, Adulterations.

664 Foods

For Adulterations see 614.3, Public health. See also 641, Cookery, and 642, Table serving.

665 Oils. Gases. Candles. Lamps

See also 644.

666 Ceramics. Glass, etc.

667 Bleaching, dyeing, etc.

668 Other organic chemical industries

669 **Metallurgy and assaying**

History of metals goes here, not in 553, 622 nor 671.

670 Manufactures

Specific topics go where of most interest. These heads are for the *general* subject of metal, wood, etc., manufactures, and for such specific manufactures as are not of more interest elsewhere. But an account of a specific manufacture is commonly most useful with its own subject; e g, a steam engine is certainly made of metal, but its manufacture should go under 621.1, not 672. The character of the library often decides the place of greatest interest. An agricultural library would prefer the manufacture and everything about wood churns with the dairy, 637, while a manufacturing library might prefer it with other wood manufactures in 674. A general library will incline to the special subject, as a more minute classification is thus obtaind, and references under 670 answer inquiries for all the library has on the subject of brass or rubber or paper manufactures. Whether groupt by use or material, if sought from the other standpoint, the references show the resources promptly. As a rule the use is the better classification, being more minute and oftener wanted. But if the uses are various the general head is better; e g. bells are for scores of distinct uses, but their manufacture is a unit; but all churns are for the dairy.

671 Articles made of metals

672 Of iron and steel; stoves, cutlery, etc.

673 Of brass and bronze; bells, etc. See also 739.

(**Manufactures**)

674 Lumber and articles made of wood
675 Leather " " " " leather
676 Paper " " " " paper
677 Cotton, wool, silk, linen, etc.
678 Rubber and articles made of rubber
679 Celluloid and other

680 Mechanic trades. Amateur manuals
681 Watch and instrumentmaking
682 Blacksmithing
683 Lock and gunmaking
684 Carriage and cabinetmaking
685 Saddlery and shoemaking. Trunks
686 Bookbinding
687 Clothesmaking. Hats
688
689 Other trades

690 Building For Architectural construction, see 721.
691 Materials. Processes. Preservativs
See also 620, Engineering.
692 Plans. Specifications, etc.
693 Masonry. Plastering. Fireproofing
694 Carpentry. Joinery. Stairbuilding
695 Roofing. Slating. Tiling See 721, Roof constructions
696 Plumbing. Gas and steam fitting, etc.
697 Heating and ventilation
See Sanitary engineering, 628; Domestic economy, 644.
698 Painting. Glazing. Paperhanging
699 Carbuilding

Fine arts

700 Fine arts. General works

701 Philosofy. Theories. Utility. Esthetics
702 Compends. Outlines
703 Dictionaries. Cyclopedias
704 Essays. Lectures. Addresses
705 Periodicals. Magazines. Reviews
706 Societies. Transactions. Reports
707 Education. Study and teaching
708 Art galleries and museums
709 History of art in general

710 Landscape gardening

711 Public parks
712 Private grounds. Lawns
713 Walks. Drives
714 Water. Fountains. Lakes
715 Trees. Hedges. Shrubs
See also 634, Forestry; 582, Botany.
716 Plants. Flowers
717 Arbors. Seats. Outlooks
718 Monuments. Mausoleums
719 Cemeteries See also 393, Treatment of the dead; 614, Public health

720 Architecture

721 Architectural construction
722 Ancient and oriental architecture (Pagan)
See also 571. Prehistoric archeology.
723 Medieval. (Christian. Mohammedan)
724 Modern
725 Public buildings
726 Ecclesiastic and religious
727 Educational and scientific
728 Residences
729 Architectural design and decoration

730 Sculpture

731 Materials and methods
732 Ancient
733 Greek and Roman
734 Medieval
735 Modern
736 Carving. Seals. Dies. Gems. Cameos
737 Numismatics. Coins. Medals
738 Pottery. Porcelain
739 Bronzes. Brasses. Bric-a-brac

740 Drawing. Decoration. Design

741 Free-hand. Crayon. Sketching from nature. Caricatures. Cartoons
Collections on any subject go with that subject with reference from 741.
742 Perspectiv See also 515, Descriptiv geometry.
743 Art anatomy. Life school
744 Mathematical and scientific drawing See also 515, Descriptiv geometry.
745 Ornamental design. Woven fabrics. Carpets. Wall paper, etc.
746 Art needlework. Fancy work
747 Interior decoration. Distemper. Fresco. Polychrome
748 Staind and iridescent glass For Leading, see 698.
749 Artistic furniture. Fireplaces. Frames, etc.

750 Painting

751 Materials and methods
752 Color See 535, Optics.
753 Epic. Mythic. Idealistic
754 Genre. Still life
755 Religious. Ecclesiastic
756 Historical. Battle scenes
757 Portrait
758 Landscape and marine. Animals. Flowers
759 Various schools of painting

760 Engraving

761 Wood
762 Copper and steel
763 Lithografy
764 Chromolithografy
765 Line and stipple
766 Mezzotint and aquatint
767 Etching. Dry point
768 Bank-note and machine
769 Collections of engravings

770 Photografy

771 Materials. Photografic chemistry
772 Silver processes. Daguerreotype. Talbotype. Collodion process. Ambrotype, etc. Dry-plate processes
773 Gelatin and pigment processes. Woodburytype. Carbon process. Lambertype. Autotype, etc.
774 Gelatin and printer's ink processes. Albertype. Heliotype. Artotype, etc.
775 Photo-lithografy, etc.
776 Photo-zincografy, etc.
777 Photo-engraving and Photo-electrotyping
778 Special applications See 522, Astronomy; 578, Microscopy.
779 Collections of photografs

780 Music

All heads include both the music itself and everything about it: Score, libretto, history, criticism, etc.

781 Theory of music
782 Dramatic music
783 Sacred music
See 245, Hymnology, sacred poetry for hymns alone; if with tunes they go in 783; 786, Organ; 246, Evangelistic use and eucharistic music; 264, Public worship.
784 Vocal music
785 Orchestral music
For Dramatic orchestral music, see 782; Sacred, 783.
786 Piano and organ
787 Stringd instruments
Including history, manufacture, instruction, music, etc.
788 Wind instruments
Including history, manufacture, instruction, music, etc.
789 Percussion and mechanical instruments

790 Amusements For Ethics of amusements, see 175.

791 Public entertainment

Concert, panorama circus, menagerie, summer resort, garden, rink, museum, fair, festival.

792 Theater. Pantomime. Opera See also 782, Dramatic music.

793 In door amusements

Private theatricals, tableaux, charades, dancing.

794 Games of skill

Chess, checkers, billiards, bowling, bagatel.

795 Games of chance

Cards, dice, backgammon, dominoes.

796 Out-door sports

Children's sports, athletic sports, coasting, skating, cycling, quoits, archery, croquet, lawn tennis.

797 Boating and ball

Rowing, yachting, base-ball, foot-ball, cricket, polo.

798 Horsemanship and racing

799 Fishing, hunting, target shooting

Literature

The classification of literature is by *language*, not country, except that North and South American are separated from other English. Canadian French goes with 84; Australian and Indian English with 820; South American Spanish with 860, etc.

800 Literature. General works

801 Philosofy. Theories. Literary esthetics
802 Compends. Outlines
803 Dictionaries. Cyclopedias
804 Essays. Lectures. Addresses
805 Periodicals. Magazines. Reviews
806 Societies. Transactions. Reports
807 Education. Study and teaching
808 Rhetoric. Treatises. Collections
809 History of literature Including general histories of books and knowledge; e. g. Hallam's *Literature of the middle ages.*

810 American literature Includes Canadian English.

811 American poetry
812 American drama
813 American fiction
814 American essays
815 American oratory
816 American letters
817 American satire and humor
818 American miscellany

820 English literature

821 English poetry
822 English drama
823 English fiction
824 English essays
825 English oratory For Speakers, see 808.
826 English letters
827 English satire and humor
828 English miscellany

Anecdotes, ana, epigrams, quotations, etc., if illustrativ of any special subject (e. g., biografy, history, science, art), go with that subject. If general they go in 828. For Riddles, Proverbs, see 398. The same rule applies to other languages in 838, 848, etc.

829 Anglo-Saxon literature

830 German literature
831 German poetry
832 German drama
833 German fiction
834 German essays
835 German oratory
836 German letters
837 German satire and humor
838 German miscellany See 828, note

839 **Minor Teutonic literatures**

840 French literature
841 French poetry
842 French drama
843 French fiction
844 French essays
845 French oratory
846 French letters
847 French satire and humor

849 **Provençal literature**

850 Italian literature
851 Italian poetry
852 Italian drama
853 Italian fiction
854 Italian essays
855 Italian oratory
856 Italian letters
857 Italian satire and humor
858 Italian miscellany

859 **Wallachian literature**

860 Spanish literature
861 Spanish poetry
862 Spanish drama
863 Spanish fiction
864 Spanish essays
865 Spanish oratory
866 Spanish letters
867 Spanish satire and humor
868 Spanish miscellany

869 **Portuguese literature**

870 Latin literature

871 Latin poetry in general
See Dramatic, Epic and Lyric below.
872 Latin drama
873 Latin epic poetry
874 Latin lyric poetry
875 Latin oratory
876 Latin letters
877 Latin satire and humor
878 Latin miscellany
879 **Medieval and modern Latin**

880 Greek literature

881 Greek poetry in general
See Dramatic, Epic and Lyric below.
882 Greek dramatic poetry
883 Greek epic poetry
884 Greek lyric poetry
885 Greek oratory
886 Greek letters
887 Greek satire and humor
For Comedies of Aristophanes see 882.
888 Greek miscellany
889 **Medieval and modern Greek**

890 Literature of minor languages

891 Minor Indo-European
892 Semitic
893 Hamitic
894 Scythian. Ural-Altaic. Turanian
Dravidian. Tamil. Finnish Turkish.
895 Eastern Asiatic. Chinese
896 African
Excluding 893. Hamitic; 892, Semitic, etc., included in families above
897 North American
898 South American
899 Malay-Polynesian and other

History

900 **History. General works**

901-909 all have *History in general* as their subject, but it is treated in these various *forms*. A periodical on English history goes with 942, not 905, which is only for periodicals on history in general.

901 Philosofy. Theories, etc. History of civilization

902 Compends. Chronologies. Charts. Outlines

For chronology as a science, see 529.

903 Dictionaries. Cyclopedias

904 Essays. Lectures. Addresses

905 Periodicals. Magazines. Reviews

906 Societies. Transactions. Reports

907 Education. Methods of teaching, writing, etc.

See also 371.

908 Polygrafy Collected works. Extracts, etc.

909 Universal and general modern histories

910 **Geografy and travels**

Including Topografy, Maps, Antiquities, Descriptions, etc.
For map projection, see 526. See also 310 Statistics, 390 Customs and costumes.
For Directories, Guide books, Gazetteers, etc., of special countries or geografical sections, see under those sections, 914-919.

911 Historical. Growth and changes in political divisions, etc.

912 Maps, atlases, plans of cities, etc.

913 Antiquities, archeology, of special countries

See also 220.9, Biblical antiquities; 340, Legal antiquities; 571, Prehistoric archeology; 930, Ancient history; 390, Special customs.

913.3 **Antiquities of ancient countries**

Antiquities of:

- .31 China
- .32 Egypt
- .33 Judea
- .34 India
- .35 Medo-Persia
- .36 Kelts
- .37 Rome. Italy
- .38 Greece
- .39 Minor countries

Antiquities of modern countries

Antiquities of

913.4 Europe
.5 Asia
.6 Africa
.7 North America
.8 South America
.9 Oceanica

914 Travel in Europe

.1 **Scotland**
.15 **Ireland**
.16 Ulster
.17 Connaught
.18 Leinster
.19 Munster
.2 **England**
.21 London
.29 Wales
.3 **Germany and Austria**
.31 Prussia and Northern Germany
.36 **Austria**
.37 Bohemia
.38 Poland
.39 Hungary
.4 **France**
.5 **Italy**
6 **Spain**
.69 **Portugal**
.7 **Russia**
.8 **Norway, Sweden and Denmark**
.81 Norway
.82 Christiania
.83 Christiansand. Bergen
.84 Hamar. Throndhjem. Tromso
.85 Sweden
.86 Gothland
.87 Svealand
.88 Norrland
.89 Denmark

914.9 **Minor countries of Europe**
.91 Iceland. Faroe islands
.92 Netherlands
.93 Belgium
.94 Switzerland
.95 Greece
.96 Turkey in Europe
.97 Servia. Bulgaria. Montenegro
.98 Roumania: Wallachia, Moldavia
.99 Islands of Grecian archipelago

915 Travel in Asia
.1 **China**
.2 **Japan**
.3 **Arabia**
.4 **India**
.5 **Persia**
.6 **Turkey in Asia**
.7 **Siberia**
.8 **Afghanistan, Turkestan, Baluchistan**
.9 **Farther India**

916 Travel in Africa
.1 **North Africa**
.2 **Egypt**
.3 **Abyssinia**
.4 **Morocco**
.5 **Algeria**
.6 **North Central Africa**
.7 **South Central Africa**
.8 **South Africa**
.9 **Madagascar. Mauritius**

917 Travel in North America
.1 **Canada. British America**
.2 **Mexico**
.28 **Central America**
.29 **West Indies**

917.3 **United States**

.4 **Northeastern or North Atlantic. New England**

.41 Maine

.42 New Hampshire

.43 Vermont

.44 Massachusetts

.45 Rhode Island

.46 Connecticut

.47 New York

.48 Pennsylvania

.49 New Jersey

.5 **Southeastern or South Atlantic**

.51 Delaware

.52 Maryland

.53 District of Columbia

.54 West Virginia

.55 Virginia

.56 North Carolina

.57 South Carolina

.58 Georgia

.59 Florida

.6 **South central or Gulf**

.61 Alabama

.62 Mississippi

.63 Louisiana

.64 Texas

.65

.66 Oklahoma. Indian Territory

.67 Arkansas

.68 Tennessee

.69 Kentucky

.7 **North central or Lake**

.71 Ohio

.72 Indiana

.73 Illinois

.74 Michigan

.75 Wisconsin

.76 Minnesota

.77 Iowa

.78 Missouri

.79

917.8 **Western or Mountain**

Including general works on "the West"

.81 Kansas
.82 Nebraska
.83 South Dakota
.84 North Dakota
.85
.86 Montana
.87 Wyoming
.88 Colorado
.89 New Mexico

.9 **Pacific**

.91 Arizona
.92 Utah
.93 Nevada
.94 California
.95 Oregon
.96 Idaho
.97 Washington
.98 Alaska
.99

918 Travel in South America

.1 **Brazil**
.2 **Argentine Republic. Patagonia**
.3 **Chili**
.4 **Bolivia**
.5 **Peru**
.6 **U. S. of Colombia. New Granada. Ecuador**
.7 **Venezuela**
.8 **Guiana**
.9 **Paraguay. Uruguay**

919 Travel in Oceanica and Polar regions

.1 **Malaysia**
.2 **Sunda**
.3 **Australasia**
.4 **Australia**
.5 **New Guinea**
.6 **Polynesia**
.7 **Isolated islands**
.8 **Arctic regions**
.9 **Antarctic regions**

920 Biografy

Collectiv biografy, including biografic dictionaries, etc. is here groupt together under the main classes: e. g. collected lives of chemists or of botanists are 925; of sailors, 926; of actors, 927.

Individual biografy includes autobiografy, diaries, personal narrativs, eulogies, etc. This is markt B in place of the class number and arranged by names of persons written about. For book numbers, see p. 17.

Collectiv by subjects

920.7 **Collected biografy of women**

Collected biografy of

921 Philosofy
922 Religion. Clergy, missionaries, preachers
923 Sociology
924 Philology
925 Science
926 Useful arts
927 Fine arts
928 Literature

Historians are put with authors in 928.

929 Genealogy and heraldry
.6 **Heraldry**
.7 **Peerages, precedence, titles of honor**
.8 **Coats of arms, crests**
.9 **Flags**

930 Ancient history To A. D. 476.

931 China
932 Egypt
933 Judea
934 India
935 Medo-Persia
936 Kelts
937 Rome. Italy
938 Greece
939 Minor countries

Modern history

940 Europe

From the fall of Western Empire (Rome), A. D. 476.

941 Scotland

941.5 Ireland

.6 **Ulster**

.7 **Connaught**

.8 **Leinster**

.9 **Munster**

942 England

Period divisions

.01	Anglo-Saxon. Including prehistoric, Roman, British, Danish.	B. C. 55-A. D. 1066.
.02	Norman.	1066–1154.
.03	Plantagenet.	1154–1399.
.04	Lancaster and York.	1400–1485.
.05	Tudor.	1485–1603.
.06	Stuart.	1603–1714.
.07	Hanover.	1714–1837.
.08	Victorian.	1837–

Geografic division

.1 **Middlesex. London**

.9 **Wales**

943 Germany and Austria

943.1 **Prussia and northern Germany**

943.6 Austria

.7 **Bohemia, etc.**

.8 **Poland.** Including the history before the division.

.9 **Hungary, etc.**

944 France

945 Italy

946 Spain

.9 **Portugal**

947 Russia

948 Norway, Sweden, and Denmark
.1 **Norway**
.2 **Christiania**
.3 **Christiansand. Bergen**
.4 **Hamar. Throndhjem. Tromso**
.5 **Sweden**
.6 **Gothland**
.7 **Svealand**
.8 **Norrland**
.9 **Denmark**

949 Minor countries of Europe
.1 **Iceland. Faroe islands**
.2 **Netherlands**
.3 **Belgium**
.4 **Switzerland**
.5 **Byzantine empire and modern Greece**
.6 **Turkey in Europe**
.7 **Servia. Bulgaria. Montenegro**
.8 **Roumania: Wallachia, Moldavia**
.9 **Islands of Grecian archipelago**

960 Asia
951 China
952 Japan
953 Arabia
954 India
955 Persia
956 Turkey in Asia See also 949.6.
957 Siberia
958 Afghanistan. Turkestan. Baluchistan
959 Farther India

960 Africa
961 North Africa
962 Egypt
963 Abyssinia
964 Morocco
965 Algeria
966 North Central Africa
967 South Central Africa
968 South Africa
969 Madagascar. Mauritius

970 North America

.1 **Indians, Aborigines**; .2 Lives of Indians; .3 Special tribes: .4 Special states; .5 Government relation and treatment; .6 Special subjects, Character, Civilization, Agriculture, etc.

See 371, Education; 572, Ethnology, 497, Languages.

971 Canada. British America.

972 Mexico. Central America

.8 Central America

.9 West Indies

973 United States and Territories

See also 342, Constitutional law and history: 328, Legislativ bodies and annals; 329, Political parties.

Period divisions.

.1 Discovery. — A. D. 896–1607.
Norse, Spanish, Dutch. French, English.

.2 Colonial. — 1607–1775.
French and Indian Wars.

.3 Revolution and Confederation. — 1775–1789.

.4 Constitutional Period. For U. S. Constitution, see 432.73. — 1789–1812.
Federalists and Republicans. See 329.1.
Washington, 1789–1797; John Adams, 1797–1801; Jefferson, 1801–1809.

.5 War of 1812. — 1812–1845.
Hartford Convention. 1814.
Nullification in South Carolina. 1832.
Annexation of Texas. 1845. See also 976.4, Texas.
Madison, 1800–1817; Monroe, 1817–1825; John Quincy Adams, 1825–1829; Jackson, 1829–1837; Van Buren, 1837–1841; Harrison, 1841; Tyler, 1841–1845.

.6 War with Mexico. — 1845–1861.
The Wilmot proviso. 1846.
Compromise of 1850.
Struggle in Kansas. 1854–1859. See also 978.1, Kansas.
Polk, 1845–1849; Taylor, 1849–1850; Fillmore, 1850–1853; Pierce, 1853–1857; Buchanan, 1857–1861.

.7 Civil War. Abolition of slavery. Lincoln, 1861–1865.

.8 Later 19th century. — 1865——.
Reconstruction. Civil service reform. Spanish War.
Johnson, 1865–1869; Grant, 1869–1877; Hayes, 1877–1881; Garfield, 1881; Arthur, 1881–1885; Cleveland, 1885–1889; Harrison, 1889–1893; Cleveland, 1893–1897; McKinley, 1897–1901.

.9 20th century.
Roosevelt, 1901–1909; Taft, 1909–

974 Northeastern or North Atlantic. New England.
.1 **Maine**
.2 **New Hampshire**
.3 **Vermont**
.4 **Massachusetts**
.5 **Rhode Island**
.6 **Connecticut**
.7 **New York**
.8 **Pennsylvania**
.9 **New Jersey**

975 Southeastern or South Atlantic
Including general works on "the South."
.1 **Delaware**
.2 **Maryland**
.3 **District of Columbia**
.4 **West Virginia**
.5 **Virginia**
.6 **North Carolina**
.7 **South Carolina**
.8 **Georgia**
.9 **Florida**

976 South central or Gulf
.1 **Alabama**
.2 **Mississippi**
.3 **Louisiana**
.4 **Texas**
.5
.6 **Oklahoma. Indian Territory**
.7 **Arkansas**
.8 **Tennessee**
.9 **Kentucky**

977 North central or Lake
.1 **Ohio**
.2 **Indiana**
.3 **Illinois**
.4 **Michigan**
.5 **Wisconsin**
.6 **Minnesota**
.7 **Iowa**
.8 **Missouri**
.9

978 Western or Mountain
Including general works on "the West."
.1 **Kansas**
.2 **Nebraska**
.3 **South Dakota** Including works on Dakota as a whole.
.4 **North Dakota**
.5
.6 **Montana**
.7 **Wyoming**
.8 **Colorado**
.9 **New Mexico**

979 Pacific
.1 **Arizona**
.2 **Utah**
.3 **Nevada**
.4 **California**
.5 **Oregon**
.6 **Idaho**
.7 **Washington**
.8 **Alaska**

980 South America

981 Brazil
982 Argentine Republic. Patagonia
983 Chili
984 Bolivia
985 Peru
986 U. S. of Colombia. New Granada. Ecuador
987 Venezuela
988 Guiana
989 Paraguay. Uruguay

990 Oceanica. Polar regions

991 Malaysia
992 Sunda
993 Australasia
994 Australia
995 New Guinea
996 Polynesia
997 Isolated islands
998 Arctic regions
999 Antarctic regions

RELATIV SUBJECT INDEX

HOW TO USE THIS INDEX

Find the subject wanted in its alfabetic place in the index. The number after it is its class number and refers to the place where the topic will be found, in numerical order of class numbers, on the shelves, shelf-lists or in the subject catalog, if used. All class numbers are decimal; e. g. 914.29 travel in Wales, comes before 914.3 travel in Germany, and both of them before 915 travel in Asia. Printed labels on the shelves or drawer fronts guide readily to the class number sought.

Under this class number will be found the resources of the library on the subject desired. Other adjoining subjects may often be profitably consulted; e. g. a description of Niagara Falls is wanted and the index refers to 917.47, but much on this topic is likely also to be found in 917.4 (travel in north Atlantic states) and even in 917.3 (travel in United States).

This index contains

in a single alfabet all subjects named in the tables, together with their synonyms and such other topics as are thought most likely to be needed by libraries using this abridgment.

If nothing is found under a number, it shows that the library does not yet possess any book on that topic.

It does not include

all names of places, minerals, plants, etc.; but gives only those used in these tables and a farther selection of the most common and useful. It is not a biografic dictionary, and gives **no names of persons** as biografic headings, but only such as appear in connection with some subject like literature or history.

RELATIV SUBJECT INDEX

Abacus	arithmetic	511
Abattoirs	public health	614
	sanitary engineering	628
Abbassides	history	956
Abbeys	architecture	726
	history	271
Abbot of misrule	customs	394
Abbots	monastic orders	271
Abbreviations	literary methods	029
	shorthand	653
Abdication	constitutional history	342
	political science	321
Abduction	criminal law	343
Aberration	astronomy	522
	mental disease	616
	hospitals	362
	insanity	132
	optics	535
Ability, human, doctrin		233
Abipones	history	982
Abode of animals, geog. distrib.		591.9
	habits	591
Abolition	political science	326
	U. S. history	973.7
Abolitionists' lives		923
Aborigines	Amer. antiquities	913.7
	Amer. history	**970.1**
	ethnology	572
	prehistoric archeology	571
Abortion	ethics	173
	law	343
	obstetrics	618
Abrasions	surgery	617
Abraxas	gems	736
	Gnostics	273
Abridging	literary methods	029
	shorthand	653
Abridgments	English law	346
	U. S. law	345
Abscess	surgery	617
Absent-mindedness	philosofy	151
Absenteeism	Irish history	941.5
	sociology	333
Absolute, the	ontology	111
Absolution	sacrament	265
Absolutism	political science	321
	predestination	234
Absorption	heat	536
	light	535
	liquids	532
Abstinence	asceticism	248
	temperance	178
Abstracting	literary methods	029
Abstraction	philosofy	153
Abyssinia	geology	556.3
	history	963
	language	492
	religious history	276.3
	travel	916.3
Academic degrees and costume		378
Academies	architecture	727
	female	376
	learned general societies	**060**
	of special subject see subject	
	private	373
	public	379
	scientific	506
Academy, New		186
	Older, Plato	184
Acadia, Nova Scotia	history	971
Acalepha	zoology	593
Accession dep't, library admin.		025
Accidence	English language	425
	French "	445
	German "	435
	Greek "	485
	Italian "	455
	Latin "	475
	Spanish "	465
Accident insurance		368
Accidents	medicin	617
	protection of public	614.8
Acclimation	hygiene	613
	of animals	591
	man	573
	plants	581
Accomplices	criminal law	343
Accountability	doctrin	233
Accounts, bookkeeping		657
	library administration	025
Accumulation	capital	331
Achaian leag		938
Achromatism	optics	535
Acids	chemical technology	661
	organic chemistry	547
	theoretic "	541
Acoustics	music	781
	physics	534
Acrobats	amusements	791
Acropolis	architecture	722

Topics in black-faced type are subdivided in tables.

Acropolis	Grecian antiquities	913.38
Acrostics	English language	828
Acting	ethics	175
	theater	792
Actinism	optics	535
	photografy	771
Actinozoa	zoology	593
Actions	legal procedure	347
Actors' lives		927
Acts and resolves	Am. statutes	345
	Eng. "	346
Acts of learned societies general		**060**
	for special subjects see topic	
	legislativ bodies, U. S.	345
Acts of the Apostles		226
Acute diseases		616
Ad valorem duty	protection	337
	taxes	336
Adages	English literature	828
	folk-lore	398
Adams, J: presidency of		973.4
J: Q. "		973.5
Addition	arithmetic	511
Addresses on special topics see topic		
Adhesion	physics	539
Adirondacks	travel	917.47
Adjutant-general's reports, U. S.		353
Administration, library		025
	of medicins, therapeutics	615
	political sci.	350
Administrativ buildings, architecture		725
	law	350
	reform	351
Administrator, law		347
Admirals' lives		923
Admiralty	foreign countries	354
Admiralty Islands	history	993
Admiralty law		347
Admission	school	371
Adobe	masonry	693
Adoption	family law	347
Adulterations, chemical analysis		543
	" **technology**	**660**
	public health	614.3
Adultery	ethics	176
	law	343
Ad valorem duty	protection	337
	taxes	336
Advent	church festivals	264
	second christology	232
Adventists, second	sect	289
Adventures	biografy	923
	travels, etc.	**910**
Adversity	pauperism	339
Advertising		659

Advocates, Faculty of		347
	lives	923
Advowsons	canon law	348
Ædiles	Roman antiquities	913.37
Æolian harp	music	787
Æolic dialect	Greek language	487
Aerated bread		641
Aerial erosion, geology		551
	navigation pneumatics	533
Aerodynamics		533
Aerolites	astronomy	523
	geology	552
Aeronautics		533
Aerofytes	lichens	589
	orchids	584
Æsthetics	fine arts	701
Æthiopia	ancient history	939
Ætolian leag	Greek history	938
Affections		157
Affidavits	law	347
Affinity	chemical	541
	family law	347
Afghanistan	geology	555.8
	history	958
	religious history	275.8
	travel	915.8
Aforisms	English literature	828
	folk-lore	398
Africa	antiquities	913.6
	botany	581.9
	geology	**556**
	history	**960**
	religious history	276
	scientific travels and surveys	508.6
	statistics	316
	travel	**916**
Africa, North	geology	556.1
	history	**961**
	religious history	276.1
	travel	**916.1**
North Central, geology		556.6
	history	966
	religious history	276.6
	travel	916.6
Propria, ancient history		939
South,	geology	556.8
	history	968
	religious history	276.8
	travel	916.8
South Central, geology		556.7
	history	967
	religious history	276.7
	travel	916.7
African colonization, sociology		325

African **history** 960
languages 496
Hamitic 493
literature 896
methodist church 287
slave trade 326
Age, influence of mental characteristics 136
therapeutics 615
of animals zoology 591
man physiology 612
natural hist. of man 573
statistics 312
pupils 379
the world 550
voters 324
Aged, hospitals for 362
Agents, consular 341
geologic 551
law 347
Ages, archeologic 571
geologic 551
Agglutination philology 401
Agnosticism natural theology 211
philosofy 149
Agnostics religious heresy 273
Agrarian laws political science 333
Roman history 937
Agreements ethics 174
law 347
Agricultural botany 581
chemistry 631
classes pol. econ. 331
colleges 630
colonization 325
experiment stations 630
geology agriculture 631
econ. geol. 553
implements 630
law 347
pests agriculture 632
" zoology 591
products pol. econ. 338
societies 630
system pol. econ. 330
Agriculture 630
cooperativ 334
department of 630
Indians 970.6
rent of land 333
Aid societies charitable associations 361
cooperation 334
Aid to the injured 614.8
Aids to readers library economy 028
lib. administration 025

Air analysis chemistry 543
brakes railroad engineering 625
compressors mech. engineering 621
disinfection contagious diseases 614
ducts ventilation 697
engines mech. engineering 621
gun military engineering 623
mechanic trades 683
hygiene 613
public health 614
meteorology 551
plants lichens 589
orchids 584
pneumatics 533
pollution public health 614
pump 533
ventilation 628
Aix-la-Chapelle internat. congress 341
treaties 341
Alabama, U. S. history 976.1
travel 917.61
claims, England 942.08
international law 341
United States 973.7
Alarms, fire protection of public 614.8
signals 654
Alaska, U. S. history 979.8
travel 917.98
purchase U. S. history 973.8
Albert N'yanza travel 916.7
Albertypes 774
Albigenses French history 944
persecutions 272
sect 284
Albinism animals 591
dermatology 616
Alchemy 540
Alcohol hygiene 613
materia medica 615
Alcoholic liquors, manufacture 663
temperance 178
Alcoholism diseases 616
Alcohols organic chemistry 547
Alcoran Mohammedanism 297
Alcoves library economy 022
Aldermen local government 352
Aldine editions rare printing 094
Ales manufacture 663
Aleutian Islands history 979.8
travel 917.98
Alexandrian library 027
philosofy 186
Alfabet elementary education 372
English orthografy 421
French " 441

Topics in black-faced type are subdivided in tables.

Alfabet	German orthografy	431
	Greek "	481
	Italian "	451
	Latin "	471
	Spanish "	461
Alfabetico-clast catalogs		019
Alfabets	ornamental design	745
	philology	411
Algae		589
Algebra		512
Algebraic logic		164
Algeria, Africa,	geology	556.5
	history	965
	religious history	276.5
	travel	916.5
Algerine war		965
	U. S. history	973.5
Algiers, Africa,	history	965
	travel	916.5
Alhambra	architecture	723
	history	946
Aliens	international law	341
	naturalization	325
	owners of land, pol. econ.	333
Alimentary canal	human anat.	611
	physiology	612
Alimony	law	347
Alkalies	chemical technology	661
	chemistry	546
Alkalin earths	chemistry	546
Alkoran	Mohammedanism	297
All-fools' day	customs	394
All-Hallows' eve	customs	394
All Saints, feast of,	church festival	264
All Souls " "	" "	264
Alleghany mountains	travel	917.5
Allegiance	English private law	347
	international law	341
	naturalization	325
Allegories	English literature	828
	religious	244
Alliance, Evangelical		206
Holy	treaties	341
Alliances	international law	341
Alligation	arithmetic	511
Alligators	zoology	598.1
Allodium	land ownership	333
	law	347
Allopathy	medicin	615
Allotropism	chemistry	541
Allotment system		333
Alloys		671
Alluvium	geology	551
Almanacs, general		529
nautical		528

Almanacs, statistical		310
Alms	ethics	177
Almshouses		362
	architecture	725
	pauperism	339
Alphabet, see Alfabet		
Alpine plants	botany	581.9
	travels	914.94
Alps	history	949.4
Altar cloths	sacred ornaments	247
Altars	ecclesiology	246
Alt-azimuth	instruments	522
Altruism	ethics	171
Alumni	biografy	378
Alwato	universal language	408
Amalgamation, metallurgy		669
Amalgams	manufactures	671
Amateur manuals	**mech. trades**	680
theatricals	amusements	793
	ethics	175
Amazon river,	travel	918.1
Amazons	customs	396
Ambassadors	international law	341
	political science	327
Amber	geology	553
	jewelry	391
	mineralogy	549
Ambrosia	antiquities	913.38
	mythology	292
Ambrotype		772
Ambulances		614.8
America, Central	geology	557.2
	history	972.8
	religious hist.	277.2
	travel	917.28
discovery of		973.1
history		970
North	botany	581.9
	geology	557
	history	970
	religious history	277
	scientific travels and surveys	508.7
	statistics	317
	travel	917
Russian,	history	979.8
	travel	917.98
South,	**geology**	558
	history	980
	religious history	278
	scientific travels and surveys	508.8
	statistics	318
	travel	918
American and Eng. private law		347

Topics in black-faced type are subdivided in tables.

American antiquities 913.7
colonies history 973.2
copyright leag 655
cyclopedias 031
essays, collected 041
independence, U.S. hist. 973.3
Indians 970.1
journalism 071
languages No. Amer. 497
So. Amer. 498
literature 810
newspapers, general 071
painting 759
party 329
periodicals, general 051
philosofy 191
political parties 329
P. E. church 283
rebellion history 973.7
revolution 973.3
slavery 326
societies, general 061
statutes law 345
Americana history 970
Americanisms 427
Amharic language 492
Amiens, peace of treaties 341
Ammergau passion play theater 792
Ammonia chemical technology 661
stimulants 615
Ammonium organic chemistry 547
Ammunition military engineering 623
Amnesty 341
Amos, book of, Bible 224
Amphibians 597
Amphictyonic council, Grecian history 938
Amphitheater antiquities 913.37
architecture 722
Amputation surgery 617
Amsterdam history 949.2
travel 914.92
Amusement or play hygiene 613.7
Amusements 790
ethics 175
of laboring classes 331
Sunday 263
Ana English literature 828
Anabaptists sects 284
Anagrams English literature 828
pseudonyms 014
Analogy logic 169
theology 219
Analysis arithmetic 511
blowpipe chemistry 544
botanical 581
Analysis **chemical** 543
grammar English 425
logic 160
microscopic 578
milk, adulterations 614.3
mineralogy 549
qualitativ, chem. 544
quantitativ, chem. 545
soil hygiene 614
sounds acoustics 534
spectrum astronomy 522
" chemistry 544
" optics 535
water biol. hygiene 613
Analysts, public adulterations 614.3
Analytic chemistry 543
geometry 516
trigonometry 514
Anam, Asia history 959
travel 915.9
Anarchists socialism 335
Anatomy animals 591
art 743
human 611
laws public health 614
pathologic 616
plants 581
vivisection ethics 179
Ancestry evolution 575
genealogy 929
mental science 136
spiritual heredity 233
traductionism 129
Anchor ice occurrence in nature 551
Ancient architecture 722
countries description 913.3
customs 913.3
customs and costume 390
geografy description 911
history 930
law 349
maps 912
music 781
philosofers' lives 921
philosofies 180
science 502
sculpture 732
Ancients, council of French legislature 328
Andersonville prison U. S. history 973.7
Andes, description 918
Andorra 946
Anecdotes legal 340
miscel. Eng. lit. 828
personal, see Biografy

Topics in black-faced type are subdivided in tables.

Anecdotes religious 244
on spec. topics, see subject
Anemometer meteorology 551
Aneroid barometer physics 533
Anesthetics therapeutics 615
Angels 235
Anger ethics 179
Angevin architecture Romanesque 723
Anglican biografy 922
brotherhoods 271
church 283
creed 238
missions 266
persecutions Eng. hist. 942.05
" relig. " 272
relig. hist. of Eng. 274.2
sisterhoods 271
Anglicisms English language 427
Angling 799
Anglo-catholics 283
Anglo-Saxon history 942.01
language 429
literature 829
Anglomania customs 390
Anilin, colors, chemical technology 667
Animal chemistry 547
electricity 134
food hygiene 613
hygiene public health 614
intelligence zoology 591
kingdom 590
magnetism 134
parasites diseases 616
plagues veterinary med. 619
psychology zoology 591
trades air pollution 614
worship, non-Christian religions 290
Animalculae 593
Animals and plants differences 577
cruelty to ethics 179
diseases veterinary medicin 619
domestic 636
evolution 575
geografic distribution 591.9
insurance 368
locomotion of 591
menageries 590
mind in 591
paleontology 560
remains of prehist. arch. 571
training of domestic 636
Animism metaphysics 128
Anne, Queen architecture 724
Annealing manufactures 671
Annealing molecular physics 539
Annexation intern. law 341
Annihilation doctrins 237
natural theology 218
Annuals, statistical 310
special, see subject
Annuities insurance 368
mathematics 519
public finance 336
Annunciation, feast of ch. festival 264
Annunciators signals 654
Anodynes therapeutics 615
Anointing sacraments 265
Anonyms bibliografy 014
Ansayrii history 955
Anselm medieval philosofy 189
Ant insects 595
Antagonism of medicins, therapeutics 615
Antartic current ocean currents 551
explorations, travel 919.9
ocean oceans 551
regions geology 559.9
history 999
religious hist. 279.9
Ante-Nicene church, A. D. 100-325 281
Anthems music 783
public worship 264
religious poetry 245
Anthologies Eng. literature 820
Anthracite econ. geology 553
fuel 644
metallurgy 669
Anthropofagy cannibalism 399
Anthropoid apes mammals 599
Anthropology mind and body 130
nat. hist. of man 573
relig. doct. of man 233
Anthropometry 573
Anthropomorfism metaphysics 125
Antiburghers presbyterians 285
Anti-Christ 235
Anti-corn law leag free trade 337
Antidotes materia medica 615
Anti-federal party 329
Antifonal choral service music 783
readings pub. worship 264
Antilles history 972.9
travel 917.29
Anti Masonry 366
Antimony inorganic chemistry 546
ores 553
Antinomianism 273
Antiquarian societies 913
Antiquarians' lives 923
Antiquities Biblical 220.9

Topics in black-faced type are subdivided in tables.

Antiquities classical 913.38
countries 913
customs 390
Jewish antiquities 913.33
history 933
religion 296
legal 340
prehistoric archeology 571
see also special subjects
Antiquity of man 573
Antiseptics public health 614
therapeutics 615
Antislavery slavery 326
U. S. history 973.7
Antitrinitarians heresies 273
unitarians 288
Ants insects 595
Antwerp history 949.3
travel 914.93
Apartment and office buildings 725
houses architecture 728
domestic economy 647
Apatite economic geology 553
fertilizers 631
mineralogy 549
Apes 599
Aphelion astronomy 521
Apiary bees 638
Apocalypse 228
Apocryfa 229
Apocryfal gospels Bible 229
Apogee moon 523
tides 525
Apollo Belvedere sculpture 733
Apollo-worship 292
Apologetics 239
Apoplexy diseases 616
Apostasy heresies 273
Apostles, acts of 226
Apostles' creed 238
lives, scriptural biografy 225
Apostolic canons apocryfa 229
catholic, sects 289
church 281
constitutions 281
epistles 229
fathers 281
succession 262
Apothecary 615
Apothegms English literature 828
folk-lore 398
Apparatus manufacture 681
practical chemistry 542
see special subject
Apparitions 133
Appetites mental faculties 158
temperance 178
Apple culture 634
Applied chemistry chem. tech. 660
mechanics mech. eng. 621
science schools 607
Appliqué work 746
Appointment of civil service officers 351
teachers education 371
Apportionment representation 324
taxation 336
Apprentices, duties of ethics 174
Apprenticeship polit. econ. 331
Appropriation bills U. S. finance 336
Aquaria 590
Aquatint engraving 766
Aqueducts canals 626
Roman antiquities 913.37
waterworks 628
Aqueous erosion geology 551
rocks " 552
Aquinas medieval philosofy 189
Arabesque decoration design 745
Arabesques ornaments 747
Arabia ancient history 939
geology 555.3
modern history 953
religious history 275.3
travel 915.3
Arabic architecture 723
language 492
literature 892
numerals arithmetic 511
philosofy 189
Arabs in Spain Spanish history 946
Aramaic language 492
Arbitrary power absolute monarchy 321
Arbitration and award law 347
capital and labor 331
international ethics 172
law 341
Arbor day 715
Arboriculture forestry 634
landscape gardening 715
Arbors " " 717
Arcades architectural design 729
walls 721
Arcadia Greek anc. history 938
Archaean age geology 551
Archaisms English language 427
Archbishops ecclesiastical polity 262
lives 922
Archeology anc. description, etc. 913
Biblical 220.9
prehistoric 571

Topics in black-faced type are subdivided in tables.

Archery		796
Arches	architectural design	729
	constr.	721
	engineering	624
	triumfal, Rom. antiq.	913.37
Archimedes screw	physics	532
Archipelago, East Indian		991
	Greek, ancient hist.	939
	modern hist.	949.9
Architects, liability of	building	692
Architects' lives		927
Architectural construction		721
	decoration	729
	design	729
	drawing	744
Architecture		720
	East Indian	722
	Gothic	723
	military engineering	623
	castles	728
	modern	724
	naval engineering	623
	Norman	723
	plans	692
	warships	623
Archives, see history		
	" manuscripts	
	of special topics see topic	
Arctic currents		551
	explorations travel	919.8
	explorers' lives	923
	ocean	551
	regions geology	559.8
	history	998
	religious history	279.8
Ardent spirits	manufactures	663
	temperance	178
Areometer	specific gravity	532
Argentine Republic	geology	558.2
	history	982
	religious history	278.2
	statistics	318
	travel	918.2
Argonauts	Greek mythology	292
Argot	French language	447
	other languages	
Argument	logic	168
	from design apologetics	239
Aria	vocal music	784
Arianism	heresies	273
Aristocracy	polit. science	321
Aristotelian philosofy		185
Arithmetic		511
	elementary education	372
Arithmetic	political statistics	310
Arithmometer		510
Arizona, U. S.	history	979.1
	travel	917.91
Arkansas, U.S.	history	976.7
	travel	917.67
Armaments	ordnance	623
Armed neutrality	international law	341
Armenian architecture		723
	church	281
	language	491
	literature	891
Armies	Amer. civil war	973.7
	military engineering	623
	military science	355
Armies	war customs	399
Arminianism	continental prot. sects	284
	methodism	287
Armistice	international law	341
Armor	war costumes	399
Armord vessels	military engineering	623
Armorican language		491
Armories	architecture	725
	military engineering	623
Arms, artificial		617
	coats of heraldry	929.8
	gun-making	683
	military engineering	623
	military science	355
	war customs	399
Army	foreign countries	354
	military science	355
	officials, lives	923
	regulations	355
	Salvation	267
	United States	353
	worm agriculture	632
	zoology	591
Arondissement	French loc. gov't	352
Arrow head inscriptions		492
	heads, stone prehist. arch.	571
	war implements	399
Arrows	archery	796
Arsacidae	history	935
Arsenals	architecture	725
	military engineering	623
Arsenic	drugs	615
	inorganic chemistry	546
	medical jurisprudence	340
	metallic poisons	615
Arson	criminal law	343
Art anatomy	drawing	743
	biografy	927
	Christian, ecclesiology	246

Art **decorativ furniture** 749
education 707
embroidery 746
fine 700
galleries 708
architecture 727
history 709
industrial, useful arts 600
interior decoration 747
needlework 746
religious 246
rudiments, prehistoric 571
sacred furniture 247
schools 707
studios architecture 727
useful 600
woman in 396
Arteries anatomy 611
diseases 616
Artesian wells, sanit. engineering 628
Arthropoda zoology 595
Arthrostraca " 595
Arthurian romances 398
Articles of confederation, U. S. const. hist. 342
thirty-nine Anglican creed 238
Articulates paleontology 565
zoology 595
Artificial butter, chemical technology 664
colors " 667
eyes 617
flowers ornamental design 745
fuel chemical technology 662
ice manufacture 621
lights chemical technology 665
limbs 617
mineral waters, chem. tech. 663
stone building materials 691
chemical technology 666
teeth dentistry 617
Artillery army dep't 358
ordnance, mil. eng. 623
Artisans' lives 926
Artistic furniture 749
Artists' lives 927
Artotype photografy 774
Arts and trades 600
fine 700
biografy of 927
grafic **fine arts** 700
useful 600
biografy of 926
Arundel marbles Greek inscrip. 481
Aryan languages 491
literatures 891
Aryan races ethnology 572
Asbestos economic geology 553
Ascensions, balloon aeronautics 533
Asceticism heresy 273
personal religion 248
Asfalt economic geology 553
pavement 625
roofing 695
Asfaltum chemical technology 665
Ashango land, history 967
travel 916.7
Ashantee history 966
travel 916.6
Ashburton treaty internat. law 341
Ashes fertilizers 631
Asia antiquities 913.05
botany 581.9
geology 555
history 950
religious history 275
scientific travels and surveys 508.5
statistics 315
travel 915
Asia Minor, ancient history 939
modern history 956
travel 915.6
Asiatic cholera diseases 616
languages 491–495
eastern 495
societies 490
Ass domestic animals 636
zoology 599
Assassination criminal law 343
Assassins lives 923
Persian history 955
Assault and battery law 343
Assaults military engineering 623
Assaying 669
Assembly, legislativ 328
Assent logic 163
Assessment taxes 336
Assignats French finance 336
Assimilation of food animals 591
man 612
plants 581
Assize, courts of law 347
Associated press Amer. journalism 070
Association mental faculties 153
Associationalism philosofy 149
Associations 360
building 332
business cooperation 334
charitable 361
educational 370

Topics in black-faced type are subdivided in tables.

Associations gilds 338
labor 331
musical 780
relig. Y. M. C., etc. 267
for special object see topic
Assurance 368
Assyria ancient history 935
antiquities 913.35
architecture 722
Assyrian language 492
Asteroids astronomy 523
Astrology 133
astronomy 520
Astronomers' lives 925
Astronomic almanacs chronology 529
instruments 522
maps 524
observations 524
spec. obs'tories 522
photografy 522
physics 521
Astronomy **520**
descriptiv 523
physical 521
practical 522
sferical 522
theoretic 521
Asylum, right of legal antiquities 340
Asylums architecture 725
hygiene of 613
inebriate architec. 725
temperance 178
sociology 362
Athanasian creed 238
Atheism apologetics 239
natural theol. 211
Atheists' lives 922
Atheneum libraries 027
Athens ancient history 938
modern history 993.5
travel 914.95
Athletic clubs 613.7
sports **796**
hygiene 613.7
school hygiene 371
Atlantic ocean, physical geografy 551
travel 910
telegraf sociology 384
useful arts 654
Atlantis antiquities 913.39
mythology 292
original home of man 572
Atlases, maps, etc. 912
celestial astron. 524
school furniture 371

Atlases, etc., see also special topics
Atmosfere geologic agent 551
meteorology 551
pneumatics 533
Atmosferic currents 551
electricity 537
engines pneumatics 533
pressure " 533
railways 625
Atolls physical geology 551
Atomic theory 541
weights 541
Atomistic philosofy 182
Atoms 541
Atonement christology 232
Attendance church 264
school 371
Attorneys law 340
lives 923
Attraction, capillary liquids 532
chemical affinity 541
gravity 531
magnetism 538
Auburn prison system prisons 365
Auction catalogs bibliografy 018
Auctions book, library econ. 025
law of 347
Audifone acoustics 534
Augsburg confession, Lutheranism 284
reformation 270
Augustine medieval philosofy 189
Augustinian monks 271
Augustinism doctrins 230
Auricular confession ordinances 265
Aurora borealis 523
Australasia geology 559.3
history 993
religious history 279.3
travel 919.3
Australia botany 581.9
geology 559.4
history 994
religious history 279.4
statistics 319
travel 919.4
Austria antiquities 913.4
botany 581.9
geology 554.3
history 943.6
religious history 274.3
statistics 314
travel 914.36
Authenticity of scripture Bible 220.1
Author bibliografies, general 011
individual 012

Topics in black-faced type are subdivided in tables.

Author catalogs	bibliografy	018
Authority	church and ministerial	262
	ethics	171
Authors and publishers		655
anon.	bibliografy	014
bibliografy of individ.		012
	special classes	013
lives		928
rights of	copyright	655
	law	347
wages		655
Authorship	literary methods	029
Auto da fe	Spanish inquisition	272
Autobiografy		920
Autocracy	form of state	321
Autografs	book rarities	091
Automata, mechanical		791
	metaphysics	127
Autotypes		773
Avalanches	physical geology	551
Average	probabilities	519
Avesta, Zend	language	491
	literature	891
	religion	295
Avignon, council of, religious history		270
Avoirdupois weights	metrology	389
Axle	physics	531
Azoic rocks	stratigrafic geology	551
Aztec	architecture	722
	history	972
	language	497
Babylon	ancient history	935
	architecture	722
	language cuneiform	492
	religion	299
Baccalaureate	sermons	252
	colleges	378
Bacchanalia	social customs	394
Bacchus	mythology	292
Bachelors	celibacy	176
Backgammon	games	795
	ethics	175
Baconian philosofy		192
	empiricism	144
Bacteria	botany	589
	theory of diseases	616
	fermentation	663
Bactrian, Old	language	491
Badges	heraldry	929.8
Bagatel	amusements	794
Bagpipe		788
Bahamas	history	972.9
	travel	917.29
Bail	criminal law	343
Baking	cookery	641
Baking powder	chem. techn.	664
	adulterations	614.3
Balance machines		531
	metrology	389
	of power political science	320
	trade commerce	382
Balances, chemical		542
Baldness	dermatology	616
Balearic Islands	history	946
Balize	history	972.8
	travel	917.28
Ball playing	amusements	797
	school hygiene	371
Ballads	Danish poetry	839
	English literature	821
	song books	398
	vocal music	784
Ballet	amusements	792
	ethics	175
Ballistics	military engineering	623
Ballooning		533
	military	623
Ballot		324
Balls	dancing	793
	ethics	175
	projectils	623
Baltimore	history	975.2
	travel	917.52
Baluchistan	geology	555.8
	history	958
	religious history	275.8
	travel	915.8
Balustrades	architectural design	729
Bampton lectures		239
Bananas	agriculture	634
Banditti	biografy	923
	law	343
Banishment	criminal law	343
Banjo		787
Bank note engraving		768
notes	money	332
Bankers' lives		923
Banking		332
Bankruptcy		347
Banks and banking	law	347
	architecture	725
	cooperativ cooperation	334
	political economy	332
	savings banks and banking	332
Banners	heraldry	929.9
	sacred ornaments	247
Banquets	social customs	394
Banting	dietetics	613
Baptism	sacrament	265
Baptismal fonts	sacred furniture	247

Topics in black-faced type are subdivided in tables.

Baptismal regeneration salvation 234
Baptisms registers of 929
Baptistery sacred furniture 247
Baptists 286
lives 922
Bar **legal** 340
biografy 923
association 347
Bar-keepers' manuals chem. tech. 663
Barbarians, antiquities see country
N. American Indians 970.1
prehistoric archeology 571
S. American Indians 980
war customs 399
Barbarism vs church 261
Barbarisms English language 428
Barbary States history 961
Barbd wire manufacture 672
Barbers 391
Bards and minstrels English lit. 821
French " 841
German " 831
Italian " 851
Provençal " 849
Spanish " 861
Barebones parliament English hist. 942.06
Barley agriculture 633
brewing 663
Barns architecture 728
Barometer pneumatics 533
Barometric leveling geodesy 526
Barometry meteorology 551
Baronage heraldry 929.7
Baronetage " 929.7
Barracks architecture 725
Barrel-organs music 789
Barricade military engineering 623
Barrows prehist. arch. 571
Bars harbor engineering 627
Barter commerce 380
Bartholomew fair public customs 394
St, religious persecution 272
Base ball amusements 797
measuring geodesy 526
Bases theoretic chemistry 541
walls 721
Bashfulness ethics 179
Basilica architecture 726
Roman antiquities 913.37
Basilisk legends 398
zoology 598.1
Basins physical geografy 551
Basket work, prehistoric 571
Basle history 949.4
travel 914.94
Bas-relief sculpture 736
Bass clarinet 788
horn wind instruments 788
Bastardy law 347
Bastile French history 944
prisons 365
Bath, order of the heraldry 929.7
Bathing customs 391
Baths architecture 725
hygiene 613
practical chemistry 542
therapeutics 615
Batrachia paleontology 567
zoology 597
Bats 599
Battalion drill, military tactics 356
Batteries, electric 537
military engineering 623
Battering ram customs of war 399
Battle paintings 756
wager of legal antiquities 340
Battledoor outdoor sports 796
Battles, see history of special countries
land military science 355
naval naval " 359
Bavaria history 943
travel 914.3
Bay windows architectural design 729
Bayeux tapestry art needlework 746
Bayonet drill military sci. 356
weapons 399
Beaches, formation of 551
Beacons navigation 656
river engineering 627
Beads archeology 571
devotional 247
Beams building 694
Bear-baiting customs 394
Beards " 391
Beasts zoology 599
of burden domestic animals 636
Beatitudes Gospels 226
Beauties, selections, Eng. lit. 828
Beauty esthetics 701
personal esthetics 701
Beaux social ethics 177
Beaver animals 599
Bechuanaland 968
Bed wagon aid to injured 614.8
Bedding furniture 645
Bedouins Arabian history 953
travel among 915.3
Beds and bedding hygiene 613
Bee-keeping 638
Beef cure therapeutics 615

Beef food .641
production domes. animals 636
Beer adulterations 614.3
analysis, chem. 543
drinking ethics 178
manufacture 663
Bees insects 595
Beet sugar chemical technology 664
Beetles insects 595
Beets garden 635
Beggars charity 177
pauperism 339
Behavior 177
Beheading punishments 343
Behring's straits travels 917.98
Being, nature of ontology 111
Bel and the dragon Apocryfa 229
Belgium antiquities 913.4
botany 581.9
geology 554.9
history 949.3
religious history 274.9
travel 914.93
zoology 591.9
Belief salvation 234
theory of knowledge 121
Bell, book and candle church discipline 262
Bell-ringing 789
Bell system education 371
Belladonna materia medica 615
Belles-lettres **800**
Belligerents, rights of inter. law 341
Bells diving hydraulic eng. 627
manufacture 673
music 789
signals 654
Belting mechan. eng. 621
Bench and bar biografy 923
Benedictine monks 271
Benediction 264
Benefactors' lives 923
Beneficence ethics 177
Benefices, church clerical support 254
Benefit of clergy legal antiquities 340
societies insurance 368
Benevolence ethics 177
Benevolent institutions 361
societies 361
parish work 256
Beowulf Anglo-Saxon poetry 829
Bequests law 347
Berbers description 916.1
Bergen, Norway history 948.3
travel 914.83
Berlin history 943.1
travel 914.31
congress, international law 341
decree " " 341
Bermudas history 972.9
Bernardines monastic orders 271
Berne history 949.4
travel 914.94
Bessel's functions 517
Bessemer process metallurgy 669
Best reading 028
Beton artificial stone 666
construction masonry 693
Betrothal customs 392
Betterments law 347
Betting ethics 175
Beverages adulterations 614.3
alcoholic, hygiene 613
hygiene 613
manufacture 663
temperance 178
Bézique amusements 795
Bezoar stones medicin 615
Bhagavat Ghita Sanskrit literature 891
Bible **220**
aids to study **220**
and science apologetics 239
antiquities 220.9
Jewish 913.33
archeology 220.9
biografy 220.9
canon 220.1
chapter and verse division 220.1
chronology 220.9
classes Sunday schools 268
codices 220.4
commentaries 220.7
concordances 220.2
criticism 220.6
cyclopedias 220.3
dictionaries 220.3
evidences 220.1
exegesis 220.6
geografy ancient geografy 911
Biblical 220.9
hermeneutics 220.6
history Biblical 220.9
Jewish history 933
in public schools education 377
inspiration 220.1
legends of 398
manuscripts 220.4
parallels Gospels 226
patriarchs 220.9
polyglots 220.5

Topics in black-faced type are subdivided in tables.

Bible revision, English		220.5
science in		220.8
societies		206
Bibliografies, general		011
Bibliografers' lives		920
Bibliografy		**010**
national		015
of individual authors		012
spec. classes of "		013
forms		014
subjects		**016**
Bibliomania	bibliografy	010
Bicycle		796
Bigamy	ethics	173
	law	343
Bigotry	ethics	179
Bilious fever	diseases	616
Bill of exchange	banks and banking	332
	law	347
rights	English	342
Billiards	ethics	175
	game	794
Billon	coins and coinage	332
Bills and notes, law		347
of mortality statistics		312
sale	law	347
Bimana	mammals	599
Bimetalism	coinage	332
Binary stars	astronomy	523
Binding, book	library economy	025
	mechanic trades	686
	rare	095
Binocular vision	physiologic optics	535
Binomial theorem	algebra	512
Biogenesis	zoology	591
Biografic	**sermons**	**920**
	homiletics	252
Biografy		**920**
	miscellaneous	920
	of gipsies	397
	Indians	970.2
	slaves	326
	scriptural	220.9
Biology		**570**
Bioplasm		576
Birds		598
cage	domestic pets	636
eggs	oölogy	598
flight of	migration	591
nests of	zoology	598
paleontology		568
Birth customs		392
marks		617
obstetrics		618
rates	public health	614

Birthday customs		**392**
Births	registers of	929
	statistics	312
Bishops	ecclesiastical polity	262
	lives of	922
Bismuth	inorganic chemistry	546
Bison	zoology	599
Bitumen	economic geology	553
Bituminous coal	" "	553
Bivalves	paleontology	564
	zoology	594
Black-art	magic	133
Forest, Germany	travel	914.3
death	plague	614
hills, Dakota	travel	917.83
hole, Calcutta		954
sea	travel	914.7
Blacking	chemical technology	667
Blackmail	law	343
Blacksmithing		682
Blasfemy	ethics	179
	law	343
Blast furnace		669
Blasting	explosivs	662
	mining engineering	622
Blazonry	heraldry	929.8
Bleaching		667
Blights	agriculture	632
Blind	asylums	362
	education of	371
Blindness		617
Blizzards	meteorology	551
Block books		092
Blockade	war	341
Blood	physiology	612
poisoning	diseases	616
vessels	anatomy	611
Bloomer costume		391
Blowing engine		621
Blowpiping	apparatus	542
	mineralogy	549
	qualitativ analysis	544
Blue books	lists of officers	351
	parliamentary	328
	United States	328
glass	therapeutics	615
laws, Connecticut		345
prints	photografy	771
Blunders	English language	428
	etiquet	395
Board of health	city gov't	352
	medicin	614
	sanitary engineer'g	628
of trade	reports commerce	381
Boarding schools	**education**	**373**

Boards	timber	674
Boat building		623
racing	ethics	175
	school hygiene	371
Boating	amusements	797
Boats, life	life saving service	656
	ship building	699
Bodily strength	athletics	613.7
Bodleian library		027
Body and mind		130
care of	hygiene	613
	school hygiene	371
measurements of		573
natural and spiritual,	dogmatics	233
symmetry of	art anatomy	743
training of	hygiene	613.7
weight of	anthropometry	573
Boers	history	968
Bog iron ore		553
Bohemia	history	943.7
	travel	914.3
Bohemian brethren	sects	284
language		491
literature		891
Bohemianism	customs	394
Bohlen lectures		239
Boilers	steam engineering	621
	insurance	368
Boiling	heat	536
points	chemistry	541
Boils	diseases	616
Bokhara	history	958
Bolivia	geology	558.4
	history	984
	religious history	278.4
	travel	918.4
Bollandists	monastic orders	271
Bolting	political parties	329
Bolts	machine tools	621
Bombs	military engineer'g	623
dynamite	chem. technology	662
Bonaparte, N.,	French hist.	944
Bondage	slavery	326
Bonds and stocks		332
civil trials		347
criminal law		343
finance		336
insurance of		368
Bone caves	prehistoric archeology	571
implements,	archeology	571
Bones	descriptiv anatomy	591
	diseases of	616
	human anatomy	611
	musical instruments	789
	paleontology	566
Bones	physiology	612
	skeleton of animal	591
Bonn conference	religious history	284
Bonnets	costume	391
	toilet	646
Book auctions		655
binding	library	025
	rare	095
	trade	686
catalogs,	author	018
	clast	017
	dictionary	019
	publishers' bibliog.	015
	subject	017
clubs	self-culture	374
collecting		025
holders	office equipment	651
illustration	inserted plates	096
	typografy	655
making	typografy	655
ownership	rarities	097
plates		097
rarities		090
trade	history	655
worm	library economy	025
Book of common prayer,	episco.	264
Bookbuying	lib. administration	025
Bookkeeping,	accounts	657
	library	025
Books and reading,	professorships of	028
bibliografy		010
classification of	lib. econ.	025
commonplace	" "	029
condemd to be burnd		098
enemies of	library economy	025
guides to selection of,	lib. econ.	028
library economy		020
lists of best	library economy	028
of prints		769
preservation of,	library econ.	025
privately printed		094
prohibited		098
reading and aids		028
restricted	library economy	024
sizes of	" "	025
tariff on	" "	025
Booksellers' catalogs,	author	018
	clast	017
	dictionary	019
lives	biografy	920
Bookselling	publishing	655
Boomerang	customs of war	399
Bootmaking		685
Booty	law	341
Bordeaux	history	944

Bordeaux	travel	914.4
Border ballads, Scottish	Eng. poet.	821
ruffians, Kansas history		978.1
Bores	social ethics	177
Borneo	history	991
	travel	919.1
Boron	inorganic chemistry	546
Borough	local gov't	352
Bosnia	history	943.9
Bosphorus	travel	914.96
Boston massacre	U. S. history	973.3
Botanic chemistry		581
gardens	architecture	727
museums	botany	580
Botanists' lives	biografy	925
Botany		**580**
	agricultural	**630**
	fossil	561
	geografic distribution	581.9
	medical	615
	structural	581
	systematic	**580**
Bottom of ocean	physical geografy	551
	zoology	591.9
Boulders, drift		551
Boundaries	law	347
	national foreign rel.	327
	intern. law	341
Bounties	army	355
	protection	337
	U. S. army	353
"Bounty" mutiny, 1789	Pitcairn's island	997
Bourbons	French history	944
Bourse	stock exchange	332
Bowls	amusements	796
Boxing	athletics	796
Boycotting	laboring classes	331
Boyle lectures,	apologetics	239
Boys' religious	societies	267
Brachiopoda	paleontology	564
	zoology	594
Brachygrafy,	shorthand	653
Brahmanism		294
Brahmo Somaj.	Indian religion	299
Brain	anatomy	611
	diseases of	616
	mental derangements	132
	mental physiology	131
	physiology	612
Brakes	railroad engineering	625
Branch libraries	library economy	022
Brandy	manufacture	663
	stimulant	615
	temperance	178

Brass instruments	music	788
	manufacture	673
Brasses, monumental	ecclesiology	247
	sculpture	739
Bravery	ethics	179
Brazil	botany	581.9
	geology	558.1
	history	981
	religious history	278.1
	statistics	318
	travel	918.1
Breach of promise	law	347
Bread	adulterations	614.3
	cookery	641
	food	641
Breadstuffs	production	338
Breakwaters	harbor engineering	627
Breast wheels	engineering	621
Breasts	diseases of women	618
Breathing	animals	591
	man	612
Breathings	Greek language	481
Breech-loading guns		683
Breeding	agriculture	636
	animals	951
Brehon laws	Irish law	349
Brethren, United	sects	284
Breton language		491
literature		891
Breviaries		264
Brewd beverages	adulterations	614.3
	chem. technol.	663
Breweries	architecture	725
Brewing		663
	air pollution	614
Bribery	criminal law	343
Bric-a-brac		739
Brick clays	economic geology	553
construction	masonry	693
Bricks	building material	691
	manufacture	666
Bridal customs		392
Bridge-building		624
Bridges	administration	351
	engineering	624
	local government	352
	military engineering	623
Bridgewater treatises		215
Brief longhand	abbreviations	653
Brigandage	criminal law	343
Brigands	law	343
Bright's disease of the kidneys		616
Briticisms	philology	427
British America	history	971
	travel	917.1

British army		354
barrows	prehist. arch.	571
Burmah	history	959
	travel	915.9
colonies	polit. sci.	325
Columbia	history	971
	travel	917.1
constitution		342
empire	**history**	**942**
	travel	914.2
Guiana	history	988
	travel	918.8
India	history	954
	travel	915.4
museum		062
	library	027
navy		354
philosofers		192
statutes		346
Britons	**history**	**942**
Brittany, France	history	944
	travel	914.4
Broad church	Anglican church	283
Broadsword exercise, military science		356
Brokerage	law	347
Bromids	hypnotics	615
Bromin	inorganic chemistry	546
Bronchi	anatomy	611
Bronchitis	diseases	616
Bronze age	pre-hist. archeology	571
	manufactures	673
Bronzes		739
Brook farm	socialism	335
Brooklyn, N. Y.	history	974.7
	travel	917.47
Brooms	domestic economy	648
Brothels	ethics	176
Brotherhoods	monks	271
	parish work	255
Brown coal	economic geology	553
Brownists	congregationalists	285
Brumaire	French history	944
Brunswick, house of, English history		942.08
Brussels	history	949.3
	travel	914.93
Brutes, immortality of	zoology	591
Bryofyta	botany	588
Bryology	"	588
Bryozoa	molluscoidea	594
Bubble, Mississippi	French finance	336
South sea	stocks "	332
Buccaneers	adventures, etc.	910
	law	343
Buchanan J.	presidency of, hist.	973.6
Bucolic poetry, English poetry		821
Buddhism		294
Buddhist architecture		722
Budget	English finance	336
Buenos Ayres,	history	982
	travel	918.2
Buffalo	zoology	599
Bugle	wind instruments	788
Bugs	rhynchota	595
Buhlwork		749
Building		**690**
	architecture	**720**
	associations, cooperativ	334
	loan institutes	332
	contracts	692
	laws building	692
	protection of life	614.8
	stones econ. geol.	553
Buildings, inspection of		614.8
	library admin.	025
	architecture	727
	economy	022
	out architecture	728
	public "	725
	local government	352
	Y. M. C. A.	267
	architecture	726
Bulgaria	history	949.7
	language	491
	travel	914.97
Bull fights	amusements	791
	ethics	175
Bulletins	library admin.	025
	publications	027
Bullion	coins and coinage	332
Bulls and bears	stock exchange	332
Irish	English humor	827
papal	eccles. polity	262
	Roman church	282
Bundesrat	German empire	354
Bundling	customs	392
Bunker hill monument		718
	history	973.3
Bunsen pump	pneumatics	533
Buoys	harbor engineering	627
	protection of life	614.8
Bureaucracy	administration	351
Bureaus	cabinet making	684
Burglar alarms	signals	654
Burglary	criminal law	343
Burial	cemeteries	719
	church rituals	265
	customs	393
	public health	614
	societies cooperation	334
Burking	criminal law	343

Topics in black-faced type are subdivided in tables.

Burlesque English literature 827
Burmah history 959
Burmese architecture 722
Burning glass optics 535
the dead customs 393
Burnings customs of war 399
student life 371
Burns and scalds surgery 617
Burying grounds cemeteries 719
public health 614
Business arithmetics 511
buildings architecture 725
colleges 650
commerce sociology 380
cooperation 334
ethics 174
manuals 658
methods 658
Butchers 689
Butter adulterations 614.3
artificial chem. technology 664
dairy 637
Butterflies 595
Buttermilk adulterations 614.3
Buttons manufacture 679
Byzantine architecture 723
art 709
ecclesiology 246
empire history 945.5
Cab manufactures 684
Cabalistic art 133
Cabinet council, English 354
making 684
officers, U. S. gov't 353
work joinery 694
Cable roads city transit 388
road engineering 625
telegrafic sociology 384
useful arts 654
Cables manufactures 677
Cabs city transit 388
Cadets military schools 355
Cadmium inorganic chemistry 546
Cædmon Anglo-Saxon poetry 829
Caesar Latin literature 878
Caesars lives 923
Roman history 937
Caesium inorganic chemistry 546
Cafés architecture 725
Cage birds domestic pets 636
Cairns prehistoric archeology 571
Calcium inorganic chemistry 546
light chemical technology 665
Calculators mathematics 510
Calculus diseases 616
Calculus mathematics 517
of direction analyt. geom. 516
Caledonia history 941
Calendar 529
ecclesiastical, Anglican 264
Calendering paper manufacture 676
textil fabrics 677
Calico manufactures 677
printing, chemical technology 667
California, U. S. history 979.4
travel 917.94
Califs Mohammedanism 297
Caligrafs writing machines 652
Caligrafy writing 652
manuscripts 091
Calisthenics hygiene 613.7
school hygiene 371
Calmucks history 950
Caloric engines 621
physics 536
Calorimeter 536
Calvinism sects 284
vs free will dogmatics 234
Calvinistic confession 238
Calvinists' lives 922
Cambistry money 332
metrology 389
Cambodian architecture 722
Cambria, Wales history 942.9
Cambrian age geology 551
confession 238
Camel domestic animals 636
zoology 599
Cameos 736
Camera photografy 771
lucida microscopy 578
optics 535
obscura " 535
photografy 771
Cameralistic science 336
Cameronians sects 285
Camfor therapeutics 615
Camp life hygiene 613
meetings 269
Campaigns political 329
Campbellism sects 289
Camping out 796
Camps, disinfec. cont. disease 614
hygiene 613
Canada botany 581.9
geology 557.1
history 971
religious history 277.1
statistics 317
travel 917.1

Topics in black-faced type are subdivided in tables.

Canal engineering 626
transit sociology 386
Canary birds care of 636
zoology 598
isles history 946
Candles chemical technology 665
Candor ethics 177
Candy adulterations 614.3
making 641
Cane, sugar agriculture 633
botany 584
Cannel coal economic geology 553
Cannibalism customs of war 399
Canning of fruits, meats, etc., chem. tech. 664
Cannon military engineering 623
Canoes and canoeing 797
Canon law 348
music 781
of Scripture 220.1
Canons, ecclesiastical polity 262
Cañons rivers 551
of Colorado river descr. 917.91
Cantata dramatic music 782
sacred music 783
Canticles Bible 223
Canvas roofing building 695
Canvassing bookselling 655
Caoutchouc organic chemistry 547
rubber manufacture 678
Capacity for knowing mental facul. 151
Cape Breton history 971
travel 917.1
Verd islands history 997
travel 919.7
Capillary attraction in plants 581
of liquids 532
Capital and labor 331
punishment law 343
Capitalists' lives 923
Capitation-tax 336
Capitols architecture 725
Caps, percussion, chemical technology 662
Captivities among the Indians 970.3
Captivity prison disciplin 365
Captivs international law 341
war customs 399
Captures international law 341
Capuchins religious history 271
Car-building 699
Caravans transportation 656
Carbines military engin'g 623
Carbolic acid materia medica 615
disinfectant pub. health 614
Carbon inorganic chemistry 546
Carbon organic chem. 547
process photografy 773
Carbonari secret societies 366
Carbonated waters hygiene 613
Carbonic acid, toxicology 615
Carboniferous age geology 551
fossils 560
Card catalog library admin. 025
playing amusements 795
ethics 175
system literary methods 029
Cardiff giant humbugs 133
Cardinals ecclesiastical polity 262
lives 922
Roman church 282
Care of animals animal hygiene 614
body, hygiene 613
school hygiene 371
person customs 391
Caribbean sea travel 917.29
Caribbee islands history 972.9
Caricature drawing 741
Carillon music 789
Carlovingian cycle legend 398
Carmelites monks 271
Carnivora mammals 599
Carnivorous plants botany 581
Carol sacred music 783
Carpentry 694
Carpets designing 745
domestic economy 645
manufacture 677
Carposporeae, algae and fungi 589
Carriage making 684
Carrier pigeons domestic birds 636
Carriers law 347
Cars building 699
horse engineering 625
railroad engineering 625
Cartesianism philosofic systems 144
Carthaginian architecture 722
Carthusian monks 271
Cartografy maps, atlases, etc. 912
Cartoons 741
Cartridges military science 623
Carving dining 642
sculpture 736
Casements fortifications 623
Cases English syntax 425
law American 345
English 346
Cashmere history 954
travel 915.4
Caspian Sea travel 914.7
Castanets musical instr. 789

Topics in black-faced type are subdivided in tables.

Caste 177
Casting foundries 671
Castles architecture 728
Castor and Pollux mythology 292
Castrametation military science 355
Casts 730
Casualties, protec. from pub. health 614.8
Casuistry ethics 171
Catacomb symbols ecclesiology 246
Catacombs antiquities 913.37
cemeteries 719
symbols 246
Catalan language 449
literature 849
Catalepsy 132
Catalog dep't library economy 025
Catalogers library staff 023
Cataloging 025
Catalogs author 018
book publishers' 015
college 378
dictionary 019
of individual authors 012
countries 015
subjects 016
star maps 524
stars 523
subject 017
universal 011
Cataracts eye diseases 617
physical geografy 551
Catarrh diseases 616
Catch vocal music 784
Catechisms 238
Caterer domestic economy 641
Cathartics therapeutics 615
Cathay history 951
travel 915.1
Cathedral canons eccles. polity 262
services musical 783
system ecclesiastical 262
Cathedrals architecture 726
Catholic apostolic church 289
church 282
epistles Bible 227
Catholicism and protestantism 282
Catholics' lives biografy 922
Cats domestic animals 636
diseases of medicin 619
Cattle 636
diseases contag. hygiene 614
vet. med. 619
Caucasia history 947
Caucuses 324
Causation philosofy 122
Cause and effect philosofy 122
Causes final " 124
of contagious diseases 614
Cavalry 357
Cavatina music 784
Cave men prehistoric archeology 571
temples Indian architecture 722
Caves geology 551
natural dwellings 571
Ceilings architecture 721
arch. decoration 729
Celebes history 991
Celestial dynamics 521
sfere 522
Celibacy ethics 176
hygiene 613
of clergy 254
Cells biology 576
histology botany 581
zoology 591
Cellular confinement prisons 365
punishments 343
tissue, botany 581
Celluloid manufactures 679
Cellulose botany 581
Celtic antiquities 913.36
Druids, religion 299
language 491
Celts ancient history 936
stone archeology 571
Cements building material 691
manufacture 666
mineral, econ. geol. 553
Cemeteries 719
earth burial 393
effect on health 614
public health 614
Cemetery chapels architecture 726
Censers sacred vessels 247
Censorship of the press 323
Censuses statistics 310
Cent-gram-second system physics 530
Centenarians physiology 612
Centennial celebrations see subject
exhibition, United States 606
Center of gravity physics 531
Centipedes zoology 595
Central Africa, North geology 556.6
history 966
religious hist. 276.6
travel 916.6
South geology 556.7
history 967
religious hist. 276.7
travel 916.7

Topics in black-faced type are subdivided in tables.

Central America geology 557.2
history 972.8
religious hist. 277.2
travel 917.28
Asia history 958
travel 915.8
Pacific railroad 385
park 711
Centralization, administrativ, foreign countries 354
U. S. gov. 353
political science 320
Centrifugal blowers engineering 621
force physics 531
pumps engineering 621
Centripetal force physics 531
Cephalopoda zoology 594
Ceramic productions building 691
Ceramics fine arts 738
manufacture 666
Cereals adulterations 614.3
agriculture 633
botany 584
trade in agric. products 338
Cerebral poisons therapeutics 615
Cerebration, unconscious 153
Cerebrospinal diseases 616
Ceremonies customs 394
religious ritual 264
Cerium inorganic chemistry 546
Certainty knowledge 121
Certificates of teachers 371
Cesium inorganic chemistry 546
Cetacea zoology 599
Ceylon history 954
travel 915.4
Chalcografy, engraving 762
Chaldea ancient history 935
Chaldean architecture 722
philosofy 181
religion 299
Chaldee language 492
Chalices sacred vessels 247
Chalk lithology 552
beds economic geology 553
Challenger expedition, 1873-76 508.3
Challenges duels 394
Chalukyan architecture 722
Chamber music 785
Chambers of commerce reports 381
Chameleon zoology 598
Champagne chemical technology 663
France, history 944
Championship games amusements 790
Championship ethics 175
school hygiene 371
Champlain, lake history 974.7
travel 917.47
Chance, games of amusements 795
ethics 175
natural theology 214
probabilities 519
vs cause metaphysics 122
Chancellors' lives 923
Chancels ecclesiology 246
Chancery law 347
reports American law 345
English law 346
Chances mathematics 519
Chandeliers 749
Change of air, hygiene 613
seasons 525
Changes of blood respiration 612
Chant, Gregorian ecclesiology 246
Chants sacred music 783
Chapbooks 398
Chapels architecture 726
Chaplains ministry 262
Chapter-and-verse division Bible 220.1
Character ethics 170
formation of 374
of reading 028
Charades 793
Charcoal chemical technology 662
drawing 741
fuel 644
Charging systems libraries 025
Charitable associations 361
institutions 361
Charities and corrections 360
Charity ethics 177
political economy 361
schools 377
sisters of 271
Charlatanry quackery 133
Charlemagne, age of, European hist. 940
legends 398
Charles 1 English history 942.06
2 " " 942.06
3 German history 943
Netherlands history 949.2
Spanish " 946
Charleston, S. C. history 975.7
travel 917.57
Charters of cities 352
Chartism Eng. constitut. history 342
Chartists political party 329
Chartografy maps, atlases, etc. 912
Charts astronomy 523

Topics in black-faced type are subdivided in tables.

Charts	geografy	912
	history	902
	maps	912
of special topic see subject		
Chase	hunting	799
Chasing	sculpture	736
Chastity	ethics	176
Chateaux	architecture	728
Chattels	law	347
Chautauqua	self-education	374
Cheating	ethics	174
	student life	371
Checkers		794
Checks	banks	332
	commercial law	347
Cheers	college	371
Cheese	adulterations	614.3
	dairy	637
Chelonia	reptils	598.1
Chemical affinity		541
	analysis	**543**
	apparatus	542
	equations	541
	fire extinguishers	614.8
	industries	**660**
	jurisprudence law	340
	laboratories	542
	manipulation	542
	manufacture	661
	nomenclature	541
	physics	541
	physiology	591
	technology	**660**
Chemistry		**540**
	agricultural	631
	analytic	**543**
	and natural religion	215
	applied to the arts	**660**
	economic zoology	591
	industrial chem. tech.	**660**
	inorganic	546
	legal law	340
	medical materia med.	615
	of poisons chem. analysis	543
	toxicology	615
	organic	547
	physiologic	612
	zoology	591
Chemists' lives	biografy	925
Chess	amusements	794
	ethics	175
Chicago, Ill.	history	977.3
	travel	917.73
Chicken-pox	diseases	616
	public health	614

Chief justices' lives		923
Chilblains	skin disorders	616
Child labor	political economy	331
Childbirth	obstetrics	618
Children, asylums for	sociology	362
	care of	649
	cruelty to ethics	179
	diseases of	618
	duties of ethics	173
	education of	372
	games of out-door	796
	hospitals for	362
	moral care for	377
Children's crusade	European history	940
Chili	botany	581.9
	geology	558.3
	history	983
	religious history	278.3
	statistics	318
	travel	918.3
Chiliasm	millennium	236
Chimes of bells	music	789
Chimneys	heating	697
	masonry	693
China	antiquities	913.31
	botany	581.9
	geology	555.1
	history, ancient	931
	modern	951
	religion, Christian	275.1
	statistics	315
	travel	915.1
ware	fine arts	738
	useful arts	666
Chinese architecture		722
	art	709
	immigration	325
	labor polit. economy	331
	language	495
	literature	895
	philosofy	181
	religion	299
Chirografy	penmanship	652
Chiromancy	palmistry	133
Chiroptera	mammals	599
Chivalry	age of Europ. hist.	940
	customs	394
Chloral	temperance	178
	therapeutics	615
Chlorin	inorganic chemistry	546
Chloroform	therapeutics	615
Chlorofyll	botany	581
Chlorosis	diseases	616
Chocolate	adulterations	614.3
	manufactures	663

Choctaw language 497
literature 897
Choice, freedom of mental faculties 159
metaphysics 123
Choir music 783
Cholera diseases 616
public health 614
Choral service sacred music 783
Chorus " " 783
vocal " 784
Chrestomathy English language 428
Christ doctrinal theology 232
lives 232
Christening baptism 265
customs 392
Christian apologetics 239
architecture eccles. 726
medieval 723
art 247
associations 267
Brothers monks 271
church 260–289
doctrins 230
education 377
ethics 171
fathers relig. hist. 270
fathers' works 281
iconografy 246
sculpt. 732
institutions 260
life personal religion 248
names 929
philosofy patristic 189
science therapeutics 615
sects 280
socialism 335
union 280
unions, young men's 267
Christiania, Norway history 948
travel 914.8
Christianity evidences 239
history 270–289
Christians biografy 922
Christiansand, Norway history 948.3
travel 914.83
Christmas customs 394
festival sacred music 783
Christology doctrinal theology 232
Chromatic aberration optics 535
scale theory of music 781
Chromatics optics 535
Chromatografy 752
Chromium economic geology 553
inorganic chem. 546
Chromolithografy 764
Chromometry colorimetry 545
Chromosfere, sun 523
Chronic diseases pathology 616
Chronicles Bible 222
history see special subject
Chronograf astronomy 522
Chronologies history 902
Chronology astronomy 529
Biblical Bible 220.9
Chronometer astron. instr. 522
longitude determ. 525
manufacture 681
Church 260–289
and clergy law 348
education 377
state eccl. polity 261
ethics 172
political science 322
Apostolic 281
architecture 726
attendance 264
authority 262
bells manufactures 673
music 789
benefices clerical support 254
Christian 260–289
decoration, sacred orn. 247
disciplin eccles. polity 262
fathers 281
fellowship 265
festivals, worship 264
furniture 247
government 262
history 270–289
membership ordinances 265
music 783
needlework 247
of England 283
Rome 282
Scotland presbyterian 285
polity 262
primitiv 270
rates clerical support 254
religious history 270–289
sacraments 265
schools 377
sects 280
services public worship 264
Slavonic language 491
sociology 322
unity 280
vestments 247
work 260
yards. cemeteries 719
Churches architecture 726

Topics in black-faced type are subdivided in tables.

Churches burials in public health 614
hygiene 613
unity of 261
Cider chemical technology 663
Cifer writing 652
Cilicia ancient history 939
modern " 956
travel 915.6
Cincinnati, O. history 977.1
travel 917.71
Cincinnati society of 369
Cinematics mechanics 531
Cinerary urns, treatment of the dead 392
Circle German local gov't 352
sailing, great navig. 527
Circuit court reports, U. S. 345
Circulating libraries 027
Circulation and blood, physiology 612
in animals zoology 591
plants botany 581
money 332
oceanic ocean currents 551
Circulatory system anatomy 611
Circumcision 392
Circumnavigations 910
Circumstantial evidence law 347
Circuses amusements 791
ethics 175
Cistercian monks 271
Cisterns rural water supply 628
Cités ouvrières, architecture 728
Cities administration 352
ancient description 913.3
buried " 913.3
free, German gov't 354
growth of, administration 352
health of 614
plans of 912
sanitation 628
Citizens, duties of ethics 172
Citizenship law 347
City governments, administration 352
halls architecture 725
houses " 728
mortality public health 614
transit sociology 388
Civil authority local government 352
engineering 620
government and the church 322
ethics 172
relig. 261
polit. science 320
law 349
liberty political science 323
lists 351
Civil pensions administration 351
procedure law 347
service administration 351
municipal gov't 352
officials, lives 923
reform admin. 351
U. S. hist. 973.8
trials 347
war ethics 172
slavery 326
U. S. history 973.7
Civility ethics 177
Civilization and the church 261
history 901
philosofy of 901
primativ 571
Claims against governments 336
U. S. court of history 347
reports 345
Clairvoyance 134
Clans political science 321
Scottish history 941
Clarinet wind instruments 788
Class feeling ethics 177
meetings public worship 264
reunions colleges 378
student customs 371
system suffrage 324
Classes, school 371
Classical antiquities 913.38
biografy 920
education 375
geografy 912
literature, Greek 880
Latin 870
mythology 292
philology Greek lang. 480
Latin " 470
schools, private, educ. 373
public " 379
Classics Greek literature 880
Latin " 870
Classification of animals zoology 590
books lib. econ. 025
chemistry 541
diseases medicin 616
knowledge method-ology 112
languages philology 401
library 025
philosofy 112
plants botany 580
sciences nat. sci. 501
Clast catalogs bibliografy 017
lib. adm. 025

Clavichord musical instruments 786
Clay chemical technology 666
petrografy 552
Clays, brick, fire, potters' geology 553
Cleanliness hygiene 613
Clearing house, banking 332
Cleavage crystallografy 548
geology 551
Clemency ethics 179
Clementines apocryfa 229
Cleopatra's needle monuments 718
Clepsydra chronology 529
Cleptomania derangements 132
Clergy, benefit of legal antiquities 340
canon law 348
celibacy of ethics 176
pastoral theol. 254
ethics 174
holy orders 262
lives of 922
pastoral work 250
Clerical support 254
taxes 336
Cleveland, G. presidency of 973.8
Cleveland, Ohio history 977.1
travel 917.71
Cliff dwellings, prehistoric arch. 571
Climacterics physiology 612
vital statistics 614
Climate, effects of diseases 616
therapeutics 615
hygiene 613
influence on man 573
meteorology 551
seasons 525
Climatology contagious diseases 614
physical geografy 551
Climbers birds 598
Climbing plants habits 581
Clinics medical education 610
Clippings literary methods 029
Clock, sidereal 522
Clock-making 681
Clocks chronology 529
Cloisterd nuns 271
Close communion baptists 286
sacrament 265
Closet, earth sanitary engineering 628
water air pollution 614
sanitary engineering 628
Cloth manufactures 677
Clothes-making 687
Clothing domestic economy 646
hygiene 613
of laboring classes 331
Clouds meteorology 551
Club foot surgery 617
houses architecture 728
libraries 027
Clubs, literary, education 374
social 367
Clusters and nebulae stars 523
Cnidaria zoology 593
Coach building carriage making 684
Coaching amusements 798
business 656
Coal, economic geology 553
fuel chemical technology 662
domestic economy 644
metallurgy 669
gas domestic lights 644
heating 697
illuminating gas 665
mineralogy 549
mining 622
oil chemical tech. 665
domestic lights 644
economic geology 553
mineralogy 549
tar colors chemical technology 667
products " " 668
Coalescence 617
Coast changes, erosion 551
defense, military eng. 623
pilot 656
survey, U. S. geodesy 526
tides 525
Coasting sports 796
Coats of arms, heraldry 929.8
Cobalt inorganic chemistry 546
ores economic geology 553
Coca hygiene 613
Cocain therapeutics 615
Cochin China, French history 959
Cochineal dyeing 667
Cock-fighting sports 791
ethics 175
Cocoa adulterations 614.3
manufacture 663
Cocoon silkworms 638
Cod fishery fishing 639
Code Napoleon French law 349
Codes, American law 345
Codices Bible mss 220.4
Codification of law U. S. law 345
Coeducation 376
of races 371
Coelenterata zoology 593
Coffee adulterations 614.3
agriculture 633

Topics in black-faced type are subdivided in tables.

Coffee botany 583
domestic economy 641
houses customs 394
hygiene 613
stimulants 615
Coffer-dams river engineering 627
Coffins burial customs 393
perishable pub. health 614
Cog wheels mech. eng. 621
Cognition metaphysics 126
Cohesion physics 539
Coils electricity 537
Coinage political economy 332
Coins numismatics 737
Coke chemical technology 662
fuel, domestic econ. 644
metallurgy 669
Cold baths hygiene 613
drinks " 613
exposure to, protection of life 614.8
physics 536
Colds catarrh 616
Colenso controversy pentateuch 222
Coleoptera 595
Colic diseases 616
Coliseum Roman antiquities 913.37
Collected biografy **920**
sermons 252
travel 910
works, see subject
Collecting biology 579
books library economy 025
Collection of taxes finance 336
Collections art galleries 708
engravings 769
of poetry Eng. lit. 821
miscel. 808
photografs 779
useful knowledge 602
vocal music 784
other topics, see subject
Collectors' manuals, biology 579
College cheers 371
degrees 378
libraries 027
regattas athletics 371
societies student life 371
songs 784
Y. M. C. A. 267
Colleges agricultural 630
and universities 378
architecture 727
business 650
for women 376
hygiene 613
Collegiants sects 284
Collegiate education women 376
Collieries mining eng. 622
Collodian process photografy 772
Colombia, So. Am. geology 558.6
history 986
religious hist. 278.6
travel 918.6
Colonial history, U. S. 973.2
law English statutes 346
system, political science 325
Colonies " " 325
Colonization, African 325
history of Africa **960**
political science 325
Colonnades architectural design 729
walls 721
Color blindness eye diseases 617
in man natural hist. of man 573
plants botany 581
of animals zoology 591
optics 535
painting 752
sense mental faculty 152
suffrage 324
Colorado, U. S., history 978.8
travel 917.88
Colord races natural hist. of man 573
Colorimetric analysis chemistry 545
Coloring dyeing 667
matters, org. chem. 547
Colors, artificial chem. tech. 667
ecclesiology 246
manufacture of paints 667
painting 752
practical printing 655
use of literary methods 029
Colosseum Roman antiquities 913.37
Colossians epistles 227
Colossus of Rhodes monuments 718
Colportage parochial 259
Columbia, District of, U. S., history 975.3
travel 917.53
Columns, architectural constr. 721
design 729
carpentry 694
sculpture 730
Combat, judicial legal antiquities 340
Combinations of capital, polit. econ. 331
labor, " " 331
Combinatory analysis algebra 512
Combustibles, storage pub. safety 614.8
Combustion chemistry 541
heat 536
Comedy English drama **822**

Topics in black-faced type are subdivided in tables.

Comets 523
maps, observations 524
theory 521
Comic journals English literature 827
music dramatic 782
works English lit. 827
Commanders' lives biografy 923
Commemorativ sermons 252
Commensalism zoology 591
Commentaries, Bible 220.7
law 347
Commerce, regulation of law 347
sociology 380
useful arts 650
Commercial agencies credit 332
arithmetics, mathematics 511
buildings, architecture 725
law 347
morals ethics 174
panics political econ. 332
traveling commerce 658
Commissions, building 692
Common carriers law 347
law, American 345
British 346
right copyright 655
treatises 347
pleas, court of 347
Prayer Anglican ritual 264
schools 379
Commonplace-books lit. methods 029
Commons, House of Eng. legis. 328
student life 371
Commonwealth English history 942.06
Commonwealths, ideal polit. science 320
Commune, French local gov't 352
German " 352
of 1871 French history 944
political economy 335
Communication, sociology 380
useful arts 650
Communion, close baptists 286
holy sacrament 265
services, sacred music 783
Communism political economy 335
Community of land political econ. 333
Compact ethics 174
Companies associations 360
joint stock law 347
Comparativ anatomy animals 591
plants 581
legislation 340
medicin 619
mythology 291
philology 410
Comparativ physiology zoology 591
religion 290
theology dogmatics 230
Compass magnetism 538
variations of, astron. 522
Compends, see subject
Competition prices 338
wages 331
Competitiv examinations, civil ser. 351
schools 371
musical festivals 784
Complex variables calculus 517
Composers' lives 927
Composition elementary English 428
music 781
rhetoric 808
type-setting 655
with creditors law 347
Compression of gases physics 533
Comprest air engines 621
Compromise bill 1850 U. S. history 973.6
ethics 171
Compulsory education 379
insurance 331
vaccination, pub. health 614
Compurgation, legal antiquities 340
Computation arithmetic 511
Concept mental philosofy 153
Conception, immaculate Christology 232
Concerts amusements 791
Conchology 594
Conclave, papal eccles. polity 262
Concord philosofy, Emerson 191
Concordances Bible 220.2
special, see subject
Concordat eccles. polity 262
internat. law 341
Concrete building material 691
construction building 693
pavements 625
roofing 695
Concretion geologic agent 551
Concubinage sex customs 392
Concussion injuries 617
Condemd books book rarities 098
Condensation action of heat 536
vapor pneumatics 533
Condensers chemistry 542
electricity 537
steam engineering 621.1
Condensing pump pneumatics 533
Condenst milk 637
Condiments adulterations 614.3
chemical technology 664
Conditional immortality 237

Topics in black-faced type are subdivided in tables.

Conduct of life ethics 170
Conductivity of rocks geology 551
Conduits telegraf 654
ventilation 697
waterworks 628
Confectionery 641
adulterations 614.3
Confederate states, U. S. civ. war 973.7
Confederation, American colonies 973.3
Conference meetings pub. worship 264
Confession and absolution, sacrament 265
Augsburg 238
auricular sacraments 265
Confessional Roman church 282
Confessions creeds 238
of faith ritual 264
Confidence practices 174
Confinement, punishment 343
schools 371
solitary, criminal law 343
prison disciplin 365
Confirmation sacraments 265
Conflagrations 614.8
Conflict of duties 171
laws, international law 341
Conformity Anglican church 283
Confucianism religion 299
Confucius Chinese philosofy 181
Congelation physics 536
Congestion of the lungs diseases 616
Conglomerate lithology 552
Congo, basin, travel 916.7
Congregation, ecclesiastical polity 262
Congregational church 285
creed 238
orders in ministry 262
singing 783
Congregationalists' lives 922
Congress of nations inter. law 341
U. S. 328
Congressional debates, U. S. 328
directories, U. S. 328
documents, U. S. 328
Conic sections analytic geom. 516
geometry 513
Conicoids analytic geometry 516
Coniferae botany 585
Conjunctions solar system 523
Conjuring sorcery 133
Connaught, Ireland history 941.7
travel 914.17
Connecticut, U. S. history 974.6
travel 917.46
Conquest international law 341
Norman, English hist. 942.02
Conscience, ethics 171
Consciousness philosofy 126
Conscription military science 355
Consecration 265
sermons 252
Conservation of energy physics 531
Conservatories, architecture 728
music 780
plants 716
Consols English finance 336
Consonants language 421
Conspiracies criminal law 343
Constables 352
Constance, council of, 1414-8, Hussites 284
religious hist. 270
Constantinople history 949.6
travel 914.96
Constellations astronomy 523
Constitution and by-laws libr. econ. 023
Constitutional conventions 342
history 342
law 342
monarchy, polit. sci. 321
Construction, architectural 721
carpentry 694
masonry 693
Consular reports 382
systems international law 341
Consuls 327
Consumption lung diseases 616
Contagious diseases pathology 616
pub. health 614
Contemplativ personal religion 248
sisterhoods 271
theology 242
Contested elections legis. bodies 328
suffrage 324
Continental protestant sects 284
travel 914
Continents physical geografy 551
Contraband of war internat. law 341
Contract system labor of convicts 331
prisons 365
Contractions shorthand 653
Contracts building 692
ethics 174
law 347
Controversial divinity 230
sermons 252
Contusions surgery 617
Conundrums amusements 793
English literature 828
Convection heat 536
Convention ecclesiastical polity 262
1792-5, French hist. 944

Topics in black-faced type are subdivided in tables.

Conventional forms decoration 729
Conventions political parties 329
Convents 271
architecture 726
Conventual education of women 376
Conversation art of 374
ethics 177
Conversion 248
revivals 269
Conveyances law 347
Convict labor political economy 331
Convicts criminal law 343
discharged, reformatory assoc. 364
Conviviality social customs 394
Convocation ecclesiastical polity 262
Convulsions diseases 616
Cook-books domestic economy 641
Cookery 641
Cooking schools 641
Coolies 326
Cooper Union, education 374
Cooperage manufactures 674
Cooperation library economy 020
literary methods 029
political economy 334
Cooperativ building associations 334
cataloging lib. admin. 025
insurance 334
Copenhagen history 948.9
travel 914.89
Copepoda articulates 595
Copernican theory astronomy 521
Copper coins money 332
engraving 762
implements archeology 571
inorganic chemistry 546
manufacture 671
metallurgy 669
ores economic geology 553
roofing 695
Coptic church sects 281
language 493
literature 893
Copts history 962
Copy and proof literary methods 029
books elementary education 372
Copying processes writing 652
Copyright international law 341
law 347
publishing 655
Coquetry ethics 177
Coral reefs geology 551
Corals zoology 593
Cordage manufacture 677
Cordials liquors 663
Cordilleras description 918
Corinthians 1 epistles 227
2 " 227
Corn agriculture 633
laws 337
trade polit. econ. 338
Cornell university 378
Cornet wind instruments 788
Cornices architec. construc. 721
design 729
Cornish language 491
Corns dermatology 616
Corona of sun 523
Coronation ceremonies 394
Coroners local government 352
Corporal punishment school 371
Corporate industries 338
Corporation, city administration 352
Corporations 360
business 380
charitable 361
law of 347
mining 622
municipal 352
railroad 385
Correction, houses of 364
Correlation of forces 531
Correspondence college, self-educ. 374
English literature 826
letter writing 808
logic 169
methods of teaching 374
natural theology 219
Swedenborgian 289
Corrosion 541
Cosmetics toilet 646
Cosmic dust astronomy 523
philosofy metaphysics 113
time chronology 529
Cosmogony metaphysics 113
natural theology 213
Cosmografy phys. geog. 551
Cosmology metaphysics 113
natural theology 213
Cossacks Russian history 947
Costume academic 378
ecclesiastic 247
hygiene 613
military 355
Costumes 391
Cottages, architecture 728
Cotton agriculture 633
dyeing chem. technol. 667
manufactures 677
trade agricultural products 338

Topics in black-faced type are subdivided in tables.

Cottonworm pests 632
Coughs and colds pathology 616
Council, ecclesiastical polity 262
general French local gov't 352
history, religious 270
municipal, Fr. local gov't 352
of state 354
privy England 354
Counsel, Christian, practical religion 241
Counterfeiting, banks and money 332
law 343
Counterpoint music 781
Country houses architecture 728
County government admin. 352
histories British 942
U. S. 974-979
Courage ethics 179
Courses of reading reading and aids 028
Court etiquet customs 395
fools " 394
Courtesy ethics 177
Courts administration 351
authority of 340
ecclesiastic 262
law 347
martial 344
Courtship ethics 177
Cousins, marriage of law 347
Coutumes, French law 349
Covenant meetings public worship 264
Scotch, creeds 238
Covenanters religious history 274.1
Scottish history 941
Covenants religious creeds 238
Covetousness ethics 179
Cow-pox diseases 616
Cowardice ethics 179
Cows domestic animals 636
Crabs zoology 595
Cramp diseases 616
Cranberries 634
Cranes and elevators mill work 621
Craniology nat. hist. of man 573
Crannoges prehistoric dwellings 571
Craters seismology 551
Crayfish zoology 595
Crayon drawing 741
Cream adulterations 614.3
Creation evolution 575
metaphysics 113
natural theology 213
Creativ power mental faculties 155
Creator doctrinal theology 231
Credibility of the Scriptures 220.1
Credit and bonds insurance 368
Credit institutes loan institutes 332
institutions administration 351
mobilier U. S. history 973.8
political economy 332
Credo sacred music 783
Creeds Anglican ritual 264
dogmatic theology 238
public worship 264
Cremation of the dead customs 393
public health 614
Crematories " " 614
Creosoting building material 691
Cresting roofs 721
Crests heraldry 929.8
Cretaceous era geology 551
fossils, paleontology 560
Crevasses glaciers 551
Crewel work art needlework 746
Cribbage games 795
Cribbing student life 371
Cricket amusements 797
Crime and illiteracy public schools 379
law 343
Crimean war English history 942.08
Russian " 947
Crimes and punishments law 343
Criminal classes reformatory assoc. 364
law 343
international 341
trials, criminal law 343
Criminals, education of 371
juvenil 343
lives 923
Crinoidea zoology 593
Crises, commercial, political economy 332
Criterion ethics 171
Critical philosofy 142
Criticism art 701
Biblical 220.7
dramatic 792
literary 801
musical 780
rhetoric 808
Croatian language 491
literature 891
Crocheting 746
Crockery pottery 738
Crocodiles 598.1
Crofton system prisons 365
Cromlechs prehistoric archeology 571
Crop insurance 368
Crops agriculture 633
Croquet 796
Cross Christian symbol 247
legends concerning 398

Topics in black-faced type are subdivided in tables.

Croton aqueduct waterworks 628
oil vegetable poisons 615
Croup diseases 616
Crozier sacred furniture 247
Crucibles metallurgy 669
practical chemistry 542
Crucifix sacred furniture 247
Cruelty ethics 179
Cruisers naval engineering 623
Cruising boating 797
Crusades European history 940
religious " 270
Crust of the earth geology 551
Crustacea paleontology 565
zoology 595
Cryptogamia 586
Cryptografic art religious art 246
Cryptografy, secret writing 652
Crystallin limestone 552
Crystallografy 548
Crystalloids 541
Cuba history 972.9
travel 917.29
Cultivated plants landscape gard. 716
Culture 301
and religion 261
self education 374
vocal 784
Cuneiform inscriptions 492
Cupping therapeutics 615
Cups archeology 571
Curativ electricity therapeutics 615
gymnastics " 615
Cures, miraculous miracles 231
therapeutics 615
Curiosa books 099
Curling amusements 796
Currency banking 332
Currents phys. geog. 551
Curriculum 375
Curvature geology 551
of spine diseases 616
Curvilinear motion physics 531
Customs and duties tarif 337
manners 390
of primitiv man 571
war 399
student life 371
taxation 336
Cutlery manufacture 672
Cuttle fish 594
Cyanogen organic chemistry 547
Cycling 796
Cyclones meteorology 551
Cyclopean remains Greek architec. 722

Cyclopedias, general 030
special, see subject
Cyfer writing 652
Cymbals musical instruments 789
Cymric languages 491
literature 891
Cynic philosofy 183
Cypher writing 652
Cypriote architecture 722
Cyrenaic philosofy 183
Daguerreotype 772
Dairy 637
products adulterations 614.3
Dais architec. accessories 729
Dakota, U. S. history 978.3
travel 917.83
Daltonism eye diseases 617
Damascus ancient history 939
travel 915.6
Damnation doctrinal theology 237
Dams mill engineering 621
mining " 622
river " 627
Dance of death emblems 249
Dances, war customs 399
Dancing amusements 793
customs 394
ethics 175
Danes English history 942.01
Dangerous classes criminal law 343
Dangers and accidents, min. eng. 622
to workmen, political econ. 331
Daniel Bible 224
Danish England English history 942.01
language 439
literature 839
Danube river travel 914.96
Danubian provinces history 949.6
travel 914.96
Darien expedition colonies 325
foreign trade 382
isthmus history 986
ship canal engineering 626
Darwinism biology 575
Dates chronology 902
Day of judgment doctrinal theology 236
schools secondary education 373
sidereal and solar 529
Days of the year chronology 529
week, names of 529
Deacon, Deaconess eccles. polity 262
Dead, disposal of the pub. health 614
ministry of sacraments 265
sea description 915.6
transportation of, public health 614

Topics in black-faced type are subdivided in tables.

Dead, treatment of the, customs 393
Deaf and dumb education of 371
institutions 362
Deafness diseases 617
Death and resurrection 236
biology 577
dance of, emblems 246
penalty, criminal law 343
physiology 612
rates public health 614
theology 236
Deaths public health 614
registers of 929
registration of, official 352
statistics 312
Debates parliamentary 328
rhetoric 808
Debating societies self-culture 374
Debt ethics 174
local government 352
public pol. econ. 336
Debtor and creditor law 347
Debts, state and national 336
Decalog Bible 222
Decay of rocks lithology 552
Decimal system arithmetic 511
weights and measures 389
Decision of character ethics 170
Decisions of juries math. prob. 519
Declamation rhetoric 808
Declaration of independence, U. S. 973.3
Savoy creeds 238
war internat. law 341
Declarations law 347
Decomposition, chemistry of 541
Decoration, architectural 729
day U. S. history 973.7
interior 747
roof architecture 721
Decorations heraldry 929.7
Decorativ art 740
Decorum ethics 177
Decrees, ecclesiastical polity 262
Decretals ecclesiastical polity 262
Roman catholic church 282
Dedication church ordinances 265
sermons 252
Deduction mental faculties 153
Deductiv logic 162
Deeds law 347
Deep boring mining eng. 622
sea dredging scientific voyages 508.3
life botany 581.9
zoology 591.9
sounding scientific voyages 508.3
Deer stalking sports 799
zoology 599
Defacements of books, library econ. 024
laws 021
Defamation law 347
Defense of religion apologetics 239
Defensiv operations, military eng. 623
Deformities surgery 617
Degeneration evolution 575
Degrees, academic 378
medical public health 614
Deism natural theology 211
Deists apologetics 239
Deity doctrinal theology 231
Delaware, U. S. history 975.1
travel 917.51
Delineation of woman in art 396
literature 396
Delirium tremens pathology 616
Deliveries and branches lib. econ. 021
Deltas physical geografy 551
Deluge Genesis 222
geologic agents 551
Delusions 133
Demagogy politics 329
Demand and supply prices 338
wages 331
Demerara history 988
travel 918.8
Democracy form of state 321
social 335
Democratic party 329
Demografy statistics 311
Demonetization 332
Demonology dogmatics 235
witchcraft 133
Dendrology botany 582
Denmark antiquities 913.4
botany 581.9
geology 554.8
history 948.9
religious history 274.8
statistics 314
travel 914.89
zoology 591.9
Denominations, Christian 280
Density gravity 531
of earth, astronomy 525
Dentistry 617
Dentists' registration, state control of 614
Denudation geology 551
Deodorization of sewage 628
Deodorizers contagious diseases 614
Departments French local gov't 352
library admin. 025

Topics in black-faced type are subdivided in tables.

Departments U. S. government 353
Deposition geology 551
Depositions evidence 347
Depravity natural theology 216
pessimism 149
total, theology 233
Depressions, industrial polit. econ. 338
Derangement, mental 132
Derivation of words, comp. philol. 412
English language 422
French " 442
German " 432
Greek " 482
Italian " 452
Latin " 472
Spanish " 462
Dermatology anatomy 611
animals 591
pathology 616
physiology 612
Dermoskeleton anatomy 591
Derricks engineering 621
Dervishes 297
Descartes empiricism 144
French philosofy 194
Descent genealogy 929
of man evolution 575
rules of law 347
Description travel **910**
Descriptiv astronomy 523
geometry 515
Deserts physical geografy 551
Design, architectural 729
arguments from 239
arts of **740**
Designing decoration 745
Desks office equipment 651
Desmidiaceae algae 589
Despotism form of state 321
Detectivs 352
Determinants algebra 512
Determinations, astron. geodesy 526
Determinativ mineralogy 549
Determinism doctrinal theology 234
philosofy 159
Detritus geology 552
Deuteronomy Bible 222
Deutsch, Platt language 439
literature 839
Developing interest in libraries 021
method of teaching 371
Development intellectual, and church 261
nutrition 591
physiology 612
theory apologetics 239
Development theory evolution 575
Deviation of the compass, magnetism 538
Devil, personal theology 235
worship 290
Devils dogmatics 235
witchcraft 133
Devonian age geology 551
fossils, paleontology **560**
Devotional exercises prayer 264
sermons 252
theology 240
Dew meteorology 551
Diabolism doctrin 235
Dialectics logic **160**
Dialects English language 427
French " 447
German " 437
Greek " 487
Italian " 457
Latin " 477
Spanish " 467
Dialogs entertainments 793
Dials chronology 529
Dialysis qualitativ analysis 544
Diamagnetism magnetism 538
Diameter of earth 525
Diamonds economic geology 553
mineralogy 549
Diaries biografy **920**
Diatessaron harmonies of gospels 226
Diathermancy heat 536
Dice ethics 175
games 795
Dicotyledonae botany 583
Dictatorship 321
Dictionaries English language 423
French " 443
German " 433
Greek " 483
Italian " 453
Latin " 473
of quotations, Eng. lit. 828
Spanish language 463
special, see subject
Dictionary catalogs 019
library admin. 025
Didactic theology 241
Die sinking 736
Diet, circle German local gov't 352
hygiene 613
Differential calculus 517
Differentiation evolution 575
Diffraction acoustics 534
optics 535
Diftheria diseases 616

Topics in black-faced type are subdivided in tables.

Diftheria public health 614
Digestion, aids to therapeutics 615
animals 591
man 612
plants 581
Digestiv system, anatomy 611
Digests American law 345
Eng. " 346
Dignities heraldry 929.7
Dikes geology 551
river and harbor eng. 627
Dilation gases 533
Diligence ethics 179
Diluvial deposits geology 551
Dimension, fourth geometry 513
Dimensions of earth 525
sun 523
Dimorfism crystallografy 548
Dining domestic economy 642
Dinners " " 642
Diocesan schools education 377
Diocese ecclesiastical polity 262
Dioptrics optics 535
Dip of rocks geology 551
Diphtheria diseases 616
public health 614
Diplomacy 327
international law 341
Diplomas school organization 371
Diplomatic hist. Amer. civil war 973.7
Diplomatics English language 421
paleografy 417
Dipsomania mental derangements 132
temperance 178
Diptera insects 595
Direct taxation, political economy 336
Directories 910
Directors library government 023
Disabilities, political suffrage 324
Disasters at sea adventures 910
navigation 656
Discharge of water engineering 627
Disciples of Christ sects 289
Disciplin, civil service 351
college education 371
ecclesiastical polity 262
education 371
mental education 370
military military science 355
prison 365
Disciplinary power administration 351
Discount banking 332
Discounts library book-buying 025
Discourses theology 252
see subject
Discoverers' lives 923
Discoveries geografy 910
science, history 509
useful arts 608
Discriminants algebra 512
Discussion, freedom of free speech 323
Disease germs, medicin 616
Diseases 616
animals veterinary med. 619
zoology 591
contagious pub. health 614
mental, medicin 616
mind and body 132
of children medicin 618
ear surgery 617
eye " 617
plants agriculture 632
botany 581
teeth dentistry 617
women pathology 618
Disfranchisement suffrage 324
Dishes pottery 738
Disinfection, contagious diseases 614
Disintegration geol. agents 551
Dislocations geology 551
surgery 617
Dispensaries institutions 362
pharmacy 615
Dispensatories materia medica 615
Dispersion optics 535
Displacement of molecules, phys. 539
Display social ethics 177
Disputations religion 204
on special subject, see subject
Dissection anatomy, animal 591
human 611
ethics 179
laws state control of 614
Dissenters English religious hist. 274.2
Dissipation ethics 178
Distance of earth from sun 525
moon and stars 523
Distances, measurement, surveying 526
Distemper painting 698
Distild beverages adulterations 614.3
chem. tech. 663
Distillation chemical technology 668
practical chemistry 542
Distortions congenital surgery 617
Distribution of animals 591.9
plants 581.9
time chronology 529
political economy 338
wealth, political economy 331
Distributiv cooperation 334

Topics in black-faced type are subdivided in tables.

District of Columbia, U. S. history 975.3
travel 917.53
court reports, U. S. 345
Dithyrambics Greek lyrics 884
Divination magic 133
Divine attributes doctrinal theol. 231
healing 265
law 231
Providence natural theol. 214
right of kings pol. sci. 321
service public worship 264
Diving bells pneumatics 533
submarine 627
Divining rod 622
Divinity schools architecture 727
theology 207
theology **230**
Divorce ethics 173
law 347
public health 614
Docks harbor engineering 627
Dockyards " " 627
Doctors' lives 926
Doctrinal sermons 252
theology **230**
Doctrins, heresies, religious hist. 273
Documents, public 328
library economy 025
Dog fighting ethics 175
tax political economy 336
Dogma **230**
Dogmas religious history 273
Dogmatic sermons 252
Dogmatics **230**
Dogs, diseases of comparativ med. 619
domestic animals 636
zoology 599
Dolmens prehist. arch. 571
Domain, eminent law 347
state finance 336
Domes architectural design 729
architecture 721
observatories astron. 522
Domesday book Eng. finance 336
history 942.02
Domestic animals 636
architecture 728
duties ethics 173
economy **640**
medicin 616
relations law 347
political sci. 323
sanitation 628
service 647
trade 381
Domestic worship, family prayer 249
Domestication, effect on animals and plants 575
Domicil law 347
Dominicans monastic orders 271
Dominoes games 795
Donatists sects 281
Donkeys domestic animals 636
Doors arch. construction 721
design 729
Doric order architecture 729
Dormer windows arch. construction 721
design 729
Dormitories architecture 727
student life 371
Dort confession 238
synod of 284
Douay Bible Eng. version 220.5
Double bass stringd instruments 787
bassoon, wind " 788
flowers floriculture 716
morfology 581
refraction optics 535
standard money 332
stars astronomy 523
Doubt metaphysics 121
Douche baths hygiene 613
Dower law 347
Drafts games 794
Dragon flies insects 595
Drainage agriculture 631
engineering 627
house 628
land 627
of cities 628
mines 622
public health 614
Drama, American literature 812
English " 822
French " 842
German " 832
Greek " 882
Italian " 852
Latin " 872
miracle plays Eng. lit. 822
gen. hist. 809
religion 244
moral influence of ethics 175
sacred 244
Spanish literature 862
Dramas Amer. civil war 973.7
Dramatic amusements 792
art theater 792
biografy actors 927
biografy Eng. authors 928

Topics in black-faced type are subdivided in tables.

Dramatic composition 808
criticism 792
history 792
music 782
orchestral music 785
Dramatists' lives, Eng. 928
Dravidian architecture 722
language 494
literature 894
Drawbridges engineering 624
Drawing **740**
architectural 744
books **740**
elementary education 372
mechanical **744**
room amusements 793
topografic 526
Drawings, architectural 729
building 692
engineering **620**
prehistoric archeology 571
Dream books folk-lore 398
Dreams 135
Dredging engineering 627
scientific voyages **508.3**
Dresden travel 914.3
Dress costume 391
ecclesiastical 247
ethics 177
Dressing hair 391
Dressmaking 646
Drift remains prehist. archeol. 571
Drifting and tunneling mining 622
Drill military science 355
Drilling mech. engineering 621
mining " 622
Drink chem. analysis 543
Drinking, social ethics 178
usages customs 394
Drinks hygiene 613
intoxicating ethics 178
Drives landscape gardening 713
Driving 798
Dropsy diseases 616
Droughts meteorology 551
Drowning protection of life 614.8
Drugs 615
adulteration of 614.3
analytic chemistry 543
economic botany 581
intoxicating ethics 178
Druids 299
Drummond light 665
Drunkenness criminal law 343
Druses religion 297
Dry docks engineering 627
goods trade production 338
plate process photografy 772
point engraving 767
rot building 691
Dualism ontology 111
Parsee heresies 273
Parseeism 295
Ducts, air ventilation 697
Dueling ancient law 340
criminal " 343
customs 394
ethics 179
Dumb asylums 362
education of 371
Dumbbells gymnastics 613.7
Dunes 551
Dunkers baptists 286
Duodecimals arithmetic 511
Duplex telegrafs communication 654
Duplicates, sale of library economy 025
Dust, air pollution 614
cosmic descriptiv astronomy 523
inhalation of hyg. of employ. 613
prevention of, in factories 628
Dutch East Indies history 991
Guiana " 988
travel 918.8
language 439
literature 839
painting 759
reformd church 284
in America 285
republic history 949.2
Duties and customs, tarif 337
moral ethics **170**
of children ethics 173
citizens " 172
husbands and wives " 173
parents " 173
public officers " 172
servants " 173
domestic econ. 647
remitted library aid 021
social ethics 177
stamp, political economy 336
Duty free import. for colleges 337
religion 233
Dwarfs biology 573
Dwelling houses architecture 728
hygiene 613
sanitation 628
Dwellings prehistoric 571
Dyadic arithmetic 511
Dyaks history 991

Topics in black-faced type are subdivided in tables.

Dyeing chem. technology 667
Dyes, anilin " " 667
Dykes engineering 627
geology 551
Dynamic electricity 537
geology 551
Dynamics 531
celestial 521
electro 537
Dynamite chemical technology 662
storage protec. of life 614.8
Dynamo-machines engineering 621.3
Dynamometer 531
Dynamos electro dynamics 537
Dysentery diseases 616
public health 614
Dyspepsia diseases 616
Ear anatomy 611
deaf and dumb asylums 362
diseases surgery 617
hospitals 362
physics 534
physiology 612
psychology 152
Early Christian architecture 723
philosofy 189
Christians history 270
education, elementary 372
English text society 820
forms of English language 427
French " 447
German " 437
Greek " 487
Italian " 457
Latin " 477
Spanish " 467
Greek philosofy 182
institutions political science 321
printed books 093
rising hygiene 613.7
Earth astronomy 525
burial of dead customs 393
pub. health 614
closet system sanitation 628
figure of 526
physical geology 551
Earthenware chemical technology 666
Earthquakes 551
Earth's attraction of moon 523
figure 526
Earthworks fortifications 623
Easements law 347
East and West, religious schism 270
India company 382
Indian archipelago, history 991
East Indian languages 491
Indies history 954
Dutch history 991
river bridge engineering 624
Easter chronology 529
church calendar 264
music 783
Eastern churches 281
empire 949.5
question, Turkish history 949.6
states U. S. history 974
Eating customs 394
diet, hygiene 613
Eccentricity in manners ethics 177
of orbit of earth 525
Eccentrics, lives of 920
machinery 621
Ecclesiastes Bible 223
Ecclesiastical antiquities 209
architecture 726
art ecclesiology 246
biografy 922
calendar chronology 529
ceremonies customs 394
costume 247
councils 262
courts 348
fasts personal relig. 248
public worship 264
feasts 264
furniture ecclesiol. 247
history 270
Latin, modern philol. 479
law canon law 348
legends 209
for special legend see topic
music 783
polity 262
statistics 209
trials 262
vestments 247
Ecclesiastics lives biografy 922
Ecclesiology church 261
symbolism 246
Echinoderms, paleontology 563
zoology 593
Eclectic medicin 615
philosofy 148
Eclecticism 148
Eclipses, finding longitude by 525
of moon or sun 523
theory 521
Ecliptic, earth 525
solar system 523
Economic botany 581

Topics in black-faced type are subdivided in tables.

Economic chemistry, chem. tech.		660
	geology	553
	zoology	591
Economics political economy		330
Economy, domestic		640
	ethics	174
	library	020
	political	330
Ecstasy	religion	248
Ecstatics	mental derangements	132
Ecuador	geology	558.6
	history	986
	religious history	278.6
	travel	918.6
Ecumenical council	eccles. polity	262
	religious hist.	270
Eden	original home of man	572
Edict of Nantes	Huguenot hist.	284
Ediles	Roman antiquities	913.37
Editions, choice of	reading and aids	028
Editorial work		070
Education		370
	and the church	261
	"	377
	in relation to labor	331
	industrial systems	371
	useful arts	607
	of voters	324
	women	376
	young	372
	self	374
	systems of	371
	useful arts	607
	special, see subject	
Educational buildings	architect.	727
	ethics	171
	reports, public	379
Educator, library as		021
Educators' lives		923
Efemerides		528
Effect and cause	metaphysics	122
Effigies, monumental		718
Effluvium nuisances	san. eng.	628
Eggs	oology	598
	poultry	636
Egina marbles, anc. sculpture		733
Egoism	ethics	171
Egypt	antiquities	913.32
	architecture	722
	botany	581.9
	geology	556.2
	hieroglyfics	493
	history, ancient	932
	modern	962
	language, old	493
Egypt	mythology	299
	obelisks monuments	718
	philosofers	181
	pyramids antiquities	913.32
	religion	299
	Soudan history	962
	travel	916.2
	travel	916.2
Egyptology	antiquities	913.32
Eight-hour law		331
Elasticity	physics	539
Elders	ecclesiastical polity	262
Eleatic Greek philosofers		182
Election	doctrinal theology	234
	frauds, U. S.	324
	sermons	252
Elections, city	administration	352
	political science	324
Electiv franchise		324
	system education	375
Electoral commission	suffrage	324
	U. S. history	973.8
	frauds	324
	system polit. science	321
Electric apparatus, inspection	protection of life	614.8
	engineering	621.3
	heating building	697
	light libraries	022
	mech. eng.	621.3
	physics	537
	manipulation chemistry	542
	measurements	537
	measuring of temperature	536
	phenom. of muscles, physiol.	612
	physics	537
	railways engineering	621.3
	shock surgery	617
	signals	537
	storms	538
	telegraf, communication	654
Electricity		537
	accidents from	614.8
	therapeutics	615
Electrobiology	animal magnetism	134
	blasting mining	622
	chronograf, astronomy	522
	dynamics	537
	magnetism	538
	metallurgy	537
	motors machines	621.3
	science	537
	physiology, animal magnetism	134
	plating metallurgy	537
Electrolysis	quantitativ analysis	545

Topics in black faced type are subdivided in tables.

Electrometers 537
Electrostatics statical 537
Electrotyping printing 655
Eleemosynary institutions 361
Elegiac poetry, English lit. 821
Greek " 884
Latin " 874
Elementary èducation 372
forms arch. design 729
readers Eng. language 428
theory of music 781
Elements, chemical chem. tech. 661
chemistry 546
Elephants mammals 599
Eleusinian mysteries Greek antiq. 913.38
mythol. 292
Elevated ground hygiene 613
railroad, city transit 388
engineering 625
temperature hygiene 613
ways public health 614
Elevation architec. design 729
Elevations meteorology 551
Elevators, mech. engineering 621
Elgin marbles 733
Elian philosofers Greek philos. 183
Eliminants algebra 512
Elixir of life alchemy 540
Elizabethan persecutions history 272
Elliptic functions calculus 517
Elocution 808
vocal culture 784
Eloquence 808
English oratory 825
Elves folk-lore 398
Elzevir editions rare printing 094
Emanation metaphysics 129
Emancipation of slaves 326
U. S. history 973.7
woman 396
Embalming of the dead customs 393
pub. health 614
Embankments, river engineering 627
Embargo 344
of 1808–12, U. S. history 973.4
Embassadors international law 341
political science 327
Emblems ecclesiology 246
Embossing, book library economy 025
Embroidery 746
Embryology animals 591
botany 581
Emergencies surgery 617
Emery wheels 621
Emigration political science 325
Emotions psychology 157
Emperors' lives 923
Empire, Eastern history 949.5
Empirical philosofy 144
Empiricism in medicin quackery 614
Employers and employd, ethics 174
polit. econ. 331
Employment agencies laboring classes 331
hygiene 613
of time ethics 179
women labor 331
customs 396
Empresses' lives 923
Emulation ethics 179
Enamel chem. technology 666
painting 751
Encaustic painting 751
Encyclical letters eccles. polity 262
Encyclopedias, general **030**
special, see subject
Encyclopedists' lives 920
Endemic diseases 616
Endemics medical geografy 614
Endless punishment future state 237
Endowd libraries 027
schools education 373
vs public schools 379
Endowment of research colleges 378
Energy, conservation of 531
Engineering, civil **620**
mechanical 621
topografic 526
Engineers' lives 926
Engines 621
England, antiquities 913.4
bank of 332
botany 581.9
church 283
geology 554.2
government 354
history **942**
language **420**
law 346
libraries 027
literature **820**
painting 759
philosofers 192
" lives 921
private law 347
religious history 274.2
sculpture 735
slavery 326
societies, general 062
statistics 314
travel 914.2

Topics in black-faced type are subdivided in tables.

England, zoology	591.9
English composition	808
cyclopedias	032
efemerides	528
essays, collected	042
journalism	072
literature	**820**
newspapers	072
periodicals, general	052
philology	**420**
private law	347
socialism	335
societies, general	062
Engravers' lives	927
occupation hygiene	613
Engraving	**760**
books of	769
of seals and gems	736
Engravings, catalogs of	016.7
collections of	769
Enigmas amusements	793
English literature	828
Enlistment army	355
Enoch, Book of	229
Enrolment school organization	371
Ensigns flags	929.9
Ensilage agriculture	633
Entail law	347
Entertaining collections, Eng. lit.	828
Entertainments, licenses for	352
public	791
Enthusiasm psychology	157
religious	269
Enthymeme logic	166
Entomology zoology	595
Entomostraca zoology	595
Entozoa parasitic diseases	616
zoology	595
Enunciation elocution	808
Environment evolution	575
influence of, therap.	615
on man	573
mental characteris.	136
Envy ethics	179
Eocene fossils paleontology	**560**
period geology	551
Eolian harp music	787
Eolic dialect	487
Ephemerides	528
Ephesians Bible	227
Epic dramatic music	782
painting	753
poetry English literature	821
Greek	883
Latin	873
Epicurean philosofy	187
Epidemics diseases	616
public health	614
Epigenesis spontan. generation	576
Epigrams English literature	828
Epilepsy diseases	616
Epiphany church festival	264
Epirus ancient history	938
Episcopacy sects	283
Episcopal charges sermons	252
church	283
orders ministry	262
Episcopalians' lives	922
Episcopate ministry	262
Epistles Bible	227
church ritual	264
Eng. literature	826
Epistolary writers' lives, English	928
Epistolografy	808
Epitafs genealogy	929
monuments	718
Epizoa parasites, zoology	591
Epizootic diseases, veterinary med.	619
Equador history	986
travel	918.6
Equality social ethics	177
Equation-machine astron. instru.	522
of center orbits	521
time chronology	529
personal, astron. corrections	522
Equations, algebraic	512
chemical	541
trigonometric	514
Equator, inclination of seasons	525
Equatorial belt earth	525
telescope astronomy	522
Equestrian exercise	798
Equilibrium astronomy, laws	521
of liquids physics	532
Equinoxes astronomy	525
Equipment of armies	355
navies	359
Equity law	347
Equivalence of force physics	531
theoretic chemistry	541
Equivocation ethics	177
Eretrian philosofers	183
Erosion	551
Erotica ethics	176
Error, sources of logic	165
Errors, grammatic Eng. language	428
of observation, probabilities	519
popular delusions	133
Erse language	491
Eruptions volcanoes	551

Erysipelas	diseases	616
Escapements	watch making	681
Eschatology		236
Esdras	Bible	229
Eskimo	history	998
Essayists' lives		928
Essays	American literature	814
	English "	824
	French "	844
	general collections	**040**
	German literature	834
	Italian "	854
	Spanish "	864
	special topics, see subject	
Essence	metaphysics	121
Essences	chem. tech.	668
Essenes	Jewish religion	296
Essential oils	chem. tech.	668
Established church, Anglican		283
Estates	law	347
of the realm		328
Esther	Apocryfa	229
	Bible	222
Esthetics	fine arts	701
Etchers' lives	biografy	927
Etching		767
Eternal life	future state	237
punishment, doctrinal theol.		237
Eternity	natural theology	218
Ether	temperance	178
	therapeutics	615
Ethers	chem. technology	668
	organic chemistry	547
Ethical culture societies		170
education		377
	ethics	170
Ethics		**170**
legal		174
of copyright		655
social		177
Ethiopia	ancient history	939
	modern "	966
Ethiopic language		492
Ethnic religions		**290**
Ethnografy		572
Ethnology		572
Etiology	medicin	616
Etiquet	customs	395
	ethics	177
Etruscan architecture		722
language		477
Etymologies	English language	422
	French "	442
	German "	432
	Greek "	482
Etymologies	Italian language	452
	Latin "	472
	Spanish "	462
Etymology, comparativ	philology	412
	inflection Eng. lang.	425
Eucharist	sacrament	265
Eucharistic music	religious art	246
vessels	sacred furniture	247
Euchre	amusements	795
Euclidian geometry		513
Eudiometer	experimental chemistry	542
Eudiometry	quantitativ analysis	545
Eufuism	rhetoric	808
Eulerian integrals	calculus	517
Eulogies	**biografy**	**920**
	sermons	252
Europe	antiquities	913.4
	botany	581.9
	geology	**554**
	history	**940**
	religious hist.	**274**
	scientific travels and surveys	508.4
	statistics	314
	travel	**914**
European politics	parties	329
Evangelical alliance	religion	206
Evangelist ministry		262
Evangelistic art		246
music	ecclesiology	246
	sacred music	783
work	pastoral theol.	253
writings, practical	"	243
Evaporation	meteorology	551
	physics	536
	practical chemistry	542
	steam engineering	621
Evening prayer	Anglican ritual	264
	family devotions	249
	personal "	248
schools		379
Evergreens	coniferae	585
Evidence	law	347
	logic	163
Evidences of Christianity		239
Evil	natural theology	216
	origin of	216
	pessimism	149
	sin	233
	spirits	235
	theodicy	231
Evolution	algebra	512
	arithmetic	511
	biology	575
	ethics	171

Topics in black-faced type are subdivided in tables.

Evolution natural theology 213
Ex libris 097
Examination, civil service 351
for academic degrees 378
of teachers, education 371
papers, English philol. 428
post mortem, pathology 616
school, education 371
Examinations, competitiv civil ser. 351
schools 371
Examining boards medical licenses 614
Excavation mining 622
Excavations, antiquities **913**
Excess of food hygiene 613
Exchange, bills of law 347
money 332
international money 332
of duplicates lib. admin. 025
specimens biology 579
stock 332
tables business 658
Exchequer, Gt. Br. finance 336
U. S. administration 353
finance 336
Excise laws 336
Excommunication, ecclesiastic polity 262
Excrementa sewerage 628
Excreta, disinfection contag. dis. 614
Excretion physiological zoology 591
Excretions, organic physiology 612
Execution criminal law 343
Executiv department const. law 342
lib. adminis. 025
Executors law 347
Exegesis Bible 220.6
Exemplaria typografica, rarities **090**
Exercise hygiene 613.7
therapeutics 615
Exertion, excessiv, effects of 616
Exhalation of vapor zoology 591
water vapor, physiology 612
Exhaustion, effects of diseases 616
Exhibition halls architecture 725
Exhibitions, building 690
international 606
of art 708
industry 606
school 371
special see subject
Exhortations public worship 264
Exile punishments 343
social ethics 177
Exodus Bible 222
Exorcism 133
Expansion heat 536
Expatriation 341
Expenditure, public finance 336
Experimental chemistry 542
Experiments, chemical 542
in special subject see subject
Exploitation mining 622
Explorations **910**
Explorers' lives 923
Explosions, history of 614.8
of fire damp, mining eng. 622
protection of life 614.8
Explosivs chem. tech. 662
protection of life 614.8
mining dangers 622
Export duties, U. S. tarif 336
Exportation 382
Exports 382
free trade 337
Expositions useful arts 606
Expository sermons 252
Exposure of children customs 392
dead " 393
to cold protection of life 614.8
Express companies sociology 385
useful arts 656
Expression, facial art anatomy 743
physiognomy 138
musical 780
Extatics mental derangements 132
Extemporaneous speaking 808
Extempore prayer public worship 264
Extension of libraries 021
teaching 374
university 378
Extinct animals paleontology **560**
plants 561
Extinction of fire protec. of life 614.8
species 575
Extra books for readers lib. econ. 024
fingers or toes deformities 617
time for readers lib. econ. 024
Extraction mining 622
of teeth dentistry 617
Extracts English literature 828
on special subjects see subject
Extradition international law 341
Extreme unction sacraments 265
Eye anatomy 611
asylums for blind 362
care of hygiene 613
diagnosis pathology 616
diseases of surgery 617
hospitals 362
optics 535
physiology 612

Topics in black-faced type are subdivided in tables.

Eye	psychology	152
Eyes, artificial	surgery	617
care of, at school		371
protection of, in factories		628
Ezekiel	Bible	224
Ezra	"	222
Fables	English literature	828
Fabrics, textil	manufacturers	677
Façades	architecture	729
Face	anatomy	611
	surgery	617
Facetiae	English literature	827
Factories	architecture	725
cooperativ	pol. econ.	334
sanitation in		628
useful arts		670
Factors and brokers	law	347
Factory, inspection	polit. econ.	331
laws	" "	338
system	labor and wages	331
Fagging	student life	371
Fahlbands	economic geology	553
Faience ware	pottery	738
Failures, trade		332
Fairies		398
Fairs, agricultural		630
amusements		791
commercial		381
industrial		606
social customs		394
Fairy tales	English literature	823
	folk-lore	398
	mythology	**291**
Faith, confessions of	creeds	238
	pub. worship	264
cure	religion	265
	therapeutics	615
doctrinal theology		234
logic of assent		163
Falconry	amusements	799
Falkland islands	history	997
	travel	919.7
Fall of man	doctrinal theology	233
Fallacies	logic	165
Fallen, parish care of		258
Falling bodies, deviations	astron.	525
laws	physics	531
Falsehood	social ethics	177
Fame	"	177
Family devotions		249
ethics		173
genealogies		929
histories		929
law		347
libraries		027
Family medicin		616
political science		321
relations	customs	392
worship		249
Famin	diseases	616
	hygiene	613
Fanaticism	religion	248
Fancy	mental faculties	155
Fancy-work		746
Fans	blowing engines	621
	mining engineering	622
	toilet	391
	ventilation	697
Fantasia	music	786
Faradization	electric medicin	615
Farces	amusements	792
	English drama	822
Farewell sermons		252
Farm-houses	architecture	728
Farmers-general	finance	336
Farmers' occupation	hygiene	613
Farming	**agriculture**	**630**
Farms, sewage,	sanitary eng.	628
Faroe islands	history	949.1
	travel	914.91
Faroese language		439
Farriery	horse shoeing	682
	vet. medicin	619
Farther India	geology	555.9
	history	959
	religious history	275.9
	travel	915.9
Fascination	animals	591
Fashion	**customs**	**390**
Fashionable education, women		376
Fast-day sermons		252
Fastening tools, mech. engineering		621
Fasti, Roman		529
Fasting	asceticism	248
	hygiene	613
	temperance	178
Fasts	church	264
Fata Morgana,	mirage	535
Fatalism	natural theol.	214
Fathers	duties	173
	of the church, hist.	270
	lives	922
	works	281
Fats	adulterations	614.3
	chem. technol.	665
	organic chemistry	547
	physiology	612
Faults	structural geology	551
	theory of, mining eng.	622
Fauna	zoology	591.9

Topics in black-faced type are subdivided in tables.

Faust, legend of legends 398
Feasts church 264
social customs 394
Feathers birds 598
comparativ anatomy 591
Fecundity physiology 612
Federal government polit. science 321
party American politics 329
Federalism political science 321
Federalists U. S. history 973.4
Feeble-minded, asylums for 362
education of 371
mental derangements 132
Feeding of animals animal hygiene 614
Feejee islands history 996
travel 919.6
Feelings emotions 157
Fees 331
Feet, artificial, surgery 617
costumes 391
English prosody 426
human anatomy 611
weights and measures 389
Fellowship, church 265
Fellowships colleges 378
Felons criminal law 343
Felt roofing 695
Female education 376
employments 396
labor polit. economy 331
sex duties 173
suffrage 324
Fences law 347
Fencing amusements 796
Fenianism Irish history 941.5
polit. assoc. 363
Fermentation chem. tech. 663
temperance 178
yeast, etc. 589
Fermented beverages, adultera. 614.3
chem. tech. 663
Ferments materia medica 615
Ferneries 716
Ferns botany 587
paleontology 561
Ferries transportation 656
Ferrotype photografy 772
Fertilization botany 581
zoology 591
Fertilizers, agriculture 631
chem. technology 668
commercial, air pollution 614
manufacture 668
from sewage 628
Festivals amusements 791
Festivals cathedral music 783
church 264
of song music 784
Fetichism 299
Feudal institutions 321
tenure law 347
Feudalism, age of, European history 940
political science 321
Fever animal heat 612
Fevers diseases 616
Fibrous materials agriculture 633
manufacture 677
Fiction American literature 813
English " 823
French " 843
German " 833
Italian " 853
reading of 028
Spanish literature 863
writing 808
Fictitious characters, dict. of 803
imprints 098
names pseudonyms 014
titles of books 098
Fidelity ethics 177
Fief 321
Field books engineering 620
fortifications 623
operations geodesy 526
sports 796
Fife musical instruments 788
Fighting, animal 175
prize 175
Figs agriculture 634
Figure-heads carving 736
human, in art art anatomy 743
of earth astronomy 525
geodesy 526
windows arch. dec. 729
Figures of heavenly bodies astron. 521
prosody Eng. language 426
French " 446
Ger. " 436
Gr. " 486
Ital. " 456
Lat. " 476
Span. " 466
Files literary methods 029
office equipment 651
Filial duty ethics 173
Filibusters criminal law 343
Cuban history 972.9
pirates 910
Filicinae botany 587
Filing machines mechanical eng. 621

Topics in black-faced type are subdivided in tables.

Filioque controversy	theology	231
Filling teeth	dentistry	617
Fillmore M., presidency of		973.6
Filth diseases,	pathology	616
	public health	614
Filtration	hygiene	613
	practical chem.	542
	sanitary eng.	628
Final causes	metaphysics	124
Finale	orchestral music	785
Finance	political economy	336
Finances	library economy	025
	local government	352
Financial admin.	civil service	351
crises	pol. economy	332
institutions	"	332
Financiers' lives		923
Finding latitude or longitude		525
	at sea	527
meridian line	astronomy	522
time		529
at sea		527
Fine arts		700
bibliografy of		016.7
biografy of		927
Fines	library economy	024
Finger alfabet, deaf and dumb		371
rings	costumes	391
Fingers, extra	deformities	617
Finishing	painting	698
Finite and infinit	metaphysics	125
differences	calculus	517
Finnish language		494
literature		894
Finno-Hungarian language		494
Fire alarm stations	architecture	725
alarms	signals	654
arms	mil. engin'g	623
	manufacture	683
balls	meteors	523
clays	economic geology	553
departments	administration	352
engines, mechanical engineering		621
	protection of life	614.8
escapes and exits		614.8
extinction	local government	352
	protection of life	614.8
insurance		368
physics		536
places		697
	artistic furniture	749
proof buildings, building		693
works	chemical tech.	662
worshipers		295
Fires	library	022
Fires	local government	352
	mines	622
	protection of life	614.8
First cause	metaphysics	122
lessons in English language		428
Fish, air pollution		614
as food	hygiene	613
commission		639
culture		639
inspection of, pub. health		614.3
preservation of	chem. tech.	664
Fisheries	business	639
	ownership, polit. econ.	333
	production	338
Fishermen's occupation	hygiene	613
Fishes	paleontology	567
	zoology	597
Fishing	sports	799
Fittings, library		022
Fixt furniture	arch. dec.	729
location	lib. economy	025
Fixtures	law	347
	library	022
Flagellants	monastic order	271
Flagellation	religious asceticism	248
Flageolet	musical instruments	788
Flags	heraldry	929.9
Flamboyant architecture, Gothic		723
Flames	physics	536
Flanders	history	949.3
Flats	architecture	728
	housekeeping	647
Flattery	ethics	177
Flax	agriculture	633
	manufacture	677
Fleets	naval science	359
Flemish language		439
literature		839
painting		759
Flies	insects	595
Flight of birds,	migration	591
Flint chips	early stone age	571
Floating batteries	naval engineering	623
bodies	physics	532
debt	finance	336
Flogging in navy		359
punishments		343
Flooding of mines		622
Floods	meteorology	551
	protection against	627
Floors and flooring	architecture	721
mosaic	arch. decoration	729
Flora, geografic		581.9
Flore and Blancheflor	legends	398
Floriculture	landscape gardening	716

Topics in black-faced type are subdivided in tables.

Florida, U. S. history 975.9
travel 917.59
Flour manufacture 679
Flow of blood physiology 612
liquids physics 532
streams meteorology 551
Flower garden, landscape garden. 716
Flowers **botany** 580
gardening 716
language of 716
Flues, smoke heating 697
Fluids, physics 532
preservativ, biol. specimens 579
Fluorescence light 535
Fluorin inorganic chemistry 546
Flute musical instruments 788
Flutina music 786
Fluxes metallurgy 669
Fluxions calculus 517
Fly, house zoology 595
Flying animal locomotion 591
machines, aeronautics 533
Fog signals commerce 656
Folds geology 551
Foliage botany 581
Folk-lore 398
song folk-lore 398
music 784
Fonoscribing shorthand 653
Font sacred furniture 247
Food, acquisition of physiolog. zool. 591
adulterations public health 614.3
animal econ. zoology 591
canned adulterations 614.3
chemical analysis 543
technology 664
cures therapeutics 615
digestion and nutrition 612
zoology 591
domestic economy 641
hygiene 613
inspection adulterations 614.3
of animals 591
laboring classes, polit. econ. 331
plants 581
preservation of 664
preservd adulterations 614.3
vegetable econ. botany 581
zoology 591
Fools lives 920
mental derangements 132
Foot English prosody 426
human anatomy 611
measures 389
prints, fossil paleontology 566
Football amusements 797
Foraminifera zoology 593
Force and energy statics 531
centrifugal and centripetal 531
electric, engineering 621.3
physics 537
intermolecular physics 539
lines of magnetism 538
metaphysics 118
of water, river engineering 627
pump pneumatics 533
transmission of, mechan. 531
mill work 621
Foreclosure law 347
Foreign banks 332
bodies, presence of, diseases 616
wounds surgery 617
elements English language 422
enlistment 355
labor political economy 331
law 349
missions 266
relations internat. law 341
polit. science 327
states administration 354
trade 382
Foreknowledge of God doct. theol. 231
Forensic chemistry law 340
medicin " 340
Foreordination salvation 234
Forest trees botany 582
Forestry agriculture 634
landscape gard. 715
Forests meteorology 551
ownership, polit. econ. 333
Forge-bellows 533
Forgemen's occupation hygiene 613
Forgery law 343
of academic degrees 378
political economy 332
Forgiveness of sin theology 234
Forks, tuning acoustics 534
Form books law 347
musical 781
ontology 111
Formation of character 374
crystals, crystallografy 548
Forms, elementary arch. design 729
Formularies materia medica 615
Fornication criminal law 343
Fortifications military engineering 623
Fortune-telling 133
Fossil man 569
antiquity of man 573
resins economic geology 553

Fossils 560
Foucault's pendulum 525
Foundation of piers bridge eng. 624
Foundations, arch. construction 721
Founding of libraries 021
type printing 655
Foundling asylums sociology 362
Foundries metal manufactures 671
Fountains landscape gard. 714
Fourierism communism 335
Fowling sport 799
Fowls domestic animals 636
Fox-hunting amusements 799
Fractional currency money 332
Fractures surgery 617
Frames artistic furniture 749
Framing carpentry 694
France antiquities 913.4
botany 581.9
geology 554.4
history 944
language 440
laws 349
literature 840
religious history 274.4
travel 914.4
zoology 591.9
Franchise, electiv polit. sci. 324
Franciscan monks 271
sisterhood 271
Franking privilege post-office 383
Franks history 943
Fraudulent elections U. S. 324
Fraunhofer's lines spectroscope 535
Free agency doctrinal theology 234
church of Scotland, sects 285
hand drawing 741
love ethics 173
news and reading rooms 027
public libraries 027
religion 211
school system education 379
soil party political party 329
U. S. history 973.6
speech polit. science 323
thinkers' lives 922
thought ecclesiastical polity 262
rationalism 211
trade and protection 337
use of libraries 024
will metaphysics 123
psychology 159
theology 234
Freebooters criminal law 343
Freedmen, education of 371
Freedmen, slavery 326
Freedmen's bureau 361
Freedom and emancipation, slavery 326
U. S. hist. 973.7
metaphysics 123
of man doct. theol. 233
press censorship 323
speech 323
political liberty 323
religious eccles. polity 262
salvation 234
Freehold land societies 334
law 347
Freemasonry 366
Freewill baptists 286
Freezing, artificial ice machines 621
physics 536
Freight transportation 656
French and Indian wars U. S. hist. 973.2
church, Gallican 282
cyclopedias 034
essays, collected 044
Guiana history 988
travel 918.8
horn wind instrum. 788
journalism 074
language 440
literature 840
newspapers 074
old language 447
opera dramatic music 782
painting 759
periodicals, general 054
philosofers' lives 921
philosofy 194
polish chem. tech. 667
protestants, persecutions of 272
revolution 944
Romanesque architecture 723
roofs arch. construc. 721
sculpture 735
societies, general 064
spoliations U. S. history 973.4
Fresco painting, building 698
interior decoration 747
painting 750
sacred ornament. 247
Fresh-air mission 361
Freshets meteorology 551
Fret-work 747
Friars monastic orders 271
Friction mech. engineering 621
physics 531
Frictional electricity 537
Friendly societies 334

Topics in black-faced type are subdivided in tables.

Friends, society of sects 289
Friendship ethics 177
Friesland history 949.2
travel 914.92
Friezes architectural decoration 729
Frisian language 439
literature 839
Fröbel's system education 372
Frogs zoology 597
Frost action geology 551
bites skin disorders 616
blight agriculture 632
meteorology 551
Fructification botany 581
Fruit culture 634
drinks adulterations 614.3
Fruits and seeds agriculture 634
organografy 581
inspec. of pub. health 614.3
preservation of chem. tech. 664
Fuel, chemical tech. 662
domestic economy 644
metallurgy 669
steam engineering 621.1
Fugitiv slave law U. S. slavery 326
Fugitivs from justice extradition 341
Fugue music 786
Fulminating powders chem. tech. 662
Fumes, prevention of, in factories 628
Fumigating apparatus, cont. diseases 614
Functional equations 517
Functions, calculus of 517
symmetric, algebra 512
trigonometric 514
Funding system finance 336
Funds, public 336
Funeral rites customs 393
religious ordinance 265
sermons 252
Fungi botany 589
Fur-trade 338
Furnaces, house warming 697
metallurgy 669
steam engineering 621.1
Furniture artistic 749
domestic economy 645
ecclesiastical 247
fixt, arch. decorat. 729
library economy 022
manufacture 684
office equipment 651
Furs clothing 646
Fuses explosivs 662
Fusion-point physics 536
Future events, probabilities 519
Future judgment doctrinal theology 236
life natural theology 218
punishment 237
state doctrinal theol. 237
Gabbro plutonic rocks 552
Gables arch. construction 721
design 729
Gadhelic language 491
Gadsden purchase U. S. history 973.6
Gaelic language 491
Gaging arithmetic 511
measures 389
Galatians Bible 227
Galena economic geology 553
metallurgy 669
Gales meteorology 551
Gallantry ethics 177
Galleries and libraries 021
art 708
architecture 727
library buildings 022
Galleys criminal law 343
Gallican church 282
Gallows criminal law 343
Galvanic manipulation chemistry 542
Galvanism 537
Galvanized iron manufactures 671
Gambling ethics 174
of amusements 175
law 345
Game laws hunting 799
Games amusements 790
champ'nship school hygiene 371
customs 394
ethics 175
Greek 913.38
of chance amusements 795
ethics 175
skill amusements 794
Roman 913.37
school 371
table and parlor 793
Garbage sanitation 628
Gardening, flower 716
kitchen 635
landscape 710
market 635
window 716
Gardens, botanic 580
public 711
zoologic 590
Garfield, J. A., presidency of 973.8
Garter, order of heraldry 929.7
Gas, coal heating 697
engines engineering **621**

Topics in black-faced type are subdivided in tables.

Gas engines physics 536
fitting 696
fuel 662
heating 697
illuminating public health 614
light domestic economy 644
library " 022
manipulation practical chem. 542
manufacture 665
natural economic geology 553
sewer air pollution 614
sanitary engin. 628
suffocation by, protection of life 614.8
water heating 697
works air pollution 614
Gases chem. tech. 665
effect on books 025
illuminating air pollution 614
in mines 622
kinetic theory 533
noxious air pollution 614
pneumatics 533
quantitativ analysis 545
Gastronomy cookery 641
Gastropoda zoology 594
Gates arch. construction 721
Gatling guns military eng. 623
Gaul history 944
Gautama Buddhism 294
Gazetteers 910
special country see country 914-919
Gearing mill engineering 621
Gehenna doctrinal theology 237
Geissler's tubes electricity 537
Gelatin processes photografy 773
Gem engraving 736
Gemara Judaism 296
Gems economic geology 553
engraved 736
mineralogy 549
Genealogy 929
General council eccles. polity 262
councils religious history 270
Generalization, logic 161
Generals, lives of 923
Generation, botany 581
spontaneous 576
steam 621.1
zoology 591
Genesis Bible 222
Geneva confession creeds 238
history 949.4
travel 914.94
tribunal international law 341
Genii 398
Genius 151
hereditary 136
Genre painting 754
Gentlemen courtesy 177
Gentleness ethics 179
Geodesy 526
Geognosy 551
Geografic coordinates 525
discovery, history of 910
distribution, animals 591.9
fossils 560
plants 581.9
names 929
Geografy, ancient 912
Biblical 220.9
descriptiv 910
elementary education 372
historical 911
maps 912
physical 551
Geologic reports 550
surveys 554-559
Geologists' lives 925
Geology 550
agricultural 631
and religion apologetics 239
applied mining eng. 622
dynamic 551
nat. theol. 215
Geomancy divination 133
Geometric drawing math. drawing 744
forms arch. decoration 729
Geometricians' lives 925
Geometry 513
analytic 516
descriptiv 515
Geomorfy 525
Georgia, U. S. history 975.8
travel 917.58
Germ theory biology 576
of disease bacteria 589
pathology 616
German catholic church sects 282
cyclopedias 033
dancing 793
empire 943
essays, collected 043
journalism 073
language 430
low 439
literature 830
low 839
mythology 293
newspapers 073
opera music 782

Topics in black-faced type are subdivided in tables.

German painting 759
periodicals, general 053
philosofers' lives 921
philosofy 193
reformation relig. hist. 274.3
secular hist. 943
reformd church 284
Romanesque architecture 723
sculpture 735
socialism 335
societies, general 063
Germany antiquities 913.4
botany 581.9
geology 554.3
government 354
history 943
laws 349
religious history 274.3
travel 914.3
zoology 591.9
Germination botany 581
Gestures oratory 808
Geysers physical geology 551
Ghebers Parseeism 295
Ghibelines Italian history 945
Ghost, Holy dogmatics 231
knights 929.7
Ghosts delusions 133
Giant, Cardiff humbugs 133
Giants natural history of man 573
Gilding 671
Gilds polit. economy 338
parish work 256
Gilead travel 915.6
Gin chemical technology 663
Gipsies biografy 397
outcast races 397
Gipsy language 397
Girders bridge engineering 624
Girls, counsels to ethics 173
female education 376
religious societies 267
Girondists, France history 944
Glacial period 551
Glaciers 551
Gladiators customs 394
Glanders diseases 616
public health 614
Glass building 691
iridescent 748
manufacture 666
painted 748
prehistoric 571
roofs architecture 721
staind, design " 729
Glazing building 698
Glees vocal music 784
Globes school apparatus 371
use of astronomy 522
Globular projection 515
Gloria in excelsis service, music 783
Glossaries English dialects 427
language 423
Gloves clothing 646
costume 391
leather, manufacture 675
woven " 677
Glucose chem. technology 664
Glue " " 668
making air pollution 614
Gluttony temperance ethics 178
Glycerin chemical technology 668
Glyfografy engraving 762
Glyptics carving 736
Gnats animal parasites 616
Gneiss lithology 552
Gnomes folk-lore 398
Gnomic poetry Greek literature 881
Gnomonic projection, descr. geom. 515
Gnosticism philosofy 189
Gnostics heresy 273
Goats 636
diseases of comparativ med. 619
Gobelin tapestry art needlework 746
Goblins folk-lore 398
God doctrins 231
Gods **idolatry** 290
mythology 291
Goiter pathology 616
Gold chemistry 546
coinage 332
manufactures 671
metallurgy 669
ores econ. geology 553
political economy 332
Golden Fleece knights 929
Goldsmithing, useful arts 671
Goldsmiths' work fine arts 739
Golf amusements 796
Goniometry crystallografy 548
Good Friday church calendar 264
music 783
Templars temperance 178
Goodness ethics 179
Goose, Mother, folk-lore 398
Gorilla zoology 599
Gospels Bible 226
church ritual 264
pseudo 229
Gossip ethics 177

Topics in black-faced type are subdivided in tables.

Gotama Buddhism 294
Gothic architecture 723
ecclesiology 246
language 439
literature 839
revival, architecture 724
Gothland, Sweden history 948.6
travel 914.86
Goths 943
Gout diseases 616
Governess, private education 373
Government central admin. 351
church 262
divine 231
elemen. schools 372
family, ethics 173
foreign countries 354
forms of 321
libraries 027
library economy 023
loans finance 336
local administration 352
political science 320
publishing 655
school, education 371
science of polit. sci. 320
U. S. and state adm. 353
Governmental buildings arch. 725
Governors' lives 923
Grace doctrinal theology 234
esthetics 701
Graces mythology 292
Graded degrees, academic 378
Grades schools 371
Grading engineering 625
Graeco-Russian church 281
Grafite economic geology 553
Graffiti Latin inscriptions 471
Grafic arts, fine arts 700
statics 531
Grafting fruit culture 634
Grail, holy legends 398
Grain agriculture 633
trade dom. trade 381
Graining house painting 698
Grains agriculture 633
Grammar, comparativ 415
elemen. education 372
schools public schools 379
universal compar. philol. 415
Grammars English language 425
French " 445
German " 435
Greek " 485
Italian " 455
Grammars Latin language 475
Spanish " 465
Grand army of the repub., association 369
U. S. hist. 973.7
opera 782
Grangers 363
Granite building rock 552
stone 553
Grant, U. S., presidency of hist. 973.8
Grants, land finance 336
to schools 379
Grape culture 634
sugar manufacture 664
Graphite economic geology 553
Grasses agriculture 633
botany 584
ornamental 716
Grasshoppers pests 632
zoology 595
Gratitude ethics 179
Grave mounds prehist. archeology 571
stones monuments 718
yards 719
Gravel disease 616
petrografy 552
Graves cemeteries 719
Gravimetric analysis chemistry 545
Gravitation celestial dynamics 521
theories 531
Gravity, center of 531
experiments geodesy 526
physics 531
specific 532
Grease chemical technology 665
Great Britain antiquities 913.4
church hist. 274.2
colonies 325
commerce 382
debt of finance 336
history 942
parliament 328
politics 329
Eastern ship-building 699
Salt lake travel 917.92
Grecian bend fashions 391
Greco-Russian church 281
Greece ancient history 938
modern botany 581.9
geology 554.9
history 949.5
travel 914.95
Greek antiquities 913.38
archipelago ancient hist. 939
modern " 949
travel 914.99

Topics in black-faced type are subdivided in tables.

Greek architecture 722
art, history of 709
church 281
classics **880**
dialects 487
fire customs of war 399
independence mod. history 949.5
language **480**
modern 489
literature **880**
modern 889
mythology 292
orders of architecture 729
paleografy inscriptions 481
philosofy **180**
revival architecture 724
sculpture 733
Greenback party 329
Greenbacks paper money 332
U. S. finance 336
Greenhouses 716
Greenland history 998
Greensand petrografy 552
Gregariousness of animals 591
Gregorian calendar 529
chant ecclesiology 246
sacred music 783
Grilles arch. construction 721
Grimm's law comparativ philology 412
Grinders' occupation hygiene 613
Grinding flour mills 679
machines mech. eng. 621
Groceries adulterations 614.3
Grocers' occupation hygiene 613
Groind vaults architecture 721
Grottos physical geografy 551
Ground air hygiene 613
public health 614
hygiene of public health 614
water physical geog. 551
Grounds, private 712
Groves landscape gardening 715
Growth biology 577
botany 581
of cities adminis. 352
statistics 312
physiology 612
zoology 591
Guano agriculture 631
economic geology 553
Guardianship law 347
Guatemala history 972.8
Guelfs Italian history 945
Guerilla warfare international law 341
Guiana, South America geology 558.8
Guiana, South America history 988
religious history 278.8
travel 918.8
Guide books **910**
Guides, railroad 656
Guilds, business political economy 338
parish work 256
Guillotin law 343
Guinea, Lower, Africa history 967
New, Oceanica " 995
Upper, Africa " 966
Guitar music 787
Gulf states, U. S. history **976**
stream physical geografy 551
Gums and resins chem. tech. 668
fossil economic geology 553
Gun cotton explosivs 662
making 683
Gunnery military engineering 623
Gunning sports 799
Gunpowder explosivs 662
manuf. protec. of life 614.8
Guns military engineering 623
Gunter's scale, mathematical instru. 510
Gutta percha rubber manufacture 678
Gymnasia German schools 379
Gymnasiums architecture 725
Gymnastics curativ 615
hygiene 613.7
school hygiene 371
vocal elocution 808
vocal music 784
Gymnosofists philosofy 181
Gymnospermae botany 585
Gynecology diseases of women 618
Gypsies biografy 397
outcast races 397
Gypsum agriculture 631
economic geology 553
lithology 552
Gypsy Romany language 397
Habakkuk Bible 224
Habeas corpus law 347
Habit mental faculties 158
Habitations, human hygiene 613
of animals zoology 591
Habits animals zoology 591
customs **390**
laboring classes, pol. econ. 331
plants physiol. botany 581
Hack licenses local government 352
Hades ancient mythology 292
doctrinal theology 236
Hæmorrhoids, diseases 616
Haggai Bible 224

Topics in black-faced type are subdivided in tables.

Hague, the history 949.2
travel 914.92
Hahnemann's theory therapeutics 615
Hail storms meteorology 551
Hair anatomy 611
diseases of 616
dressing customs 391
toilet 646
organic chemistry 547
physiology 612
Half-way covenant 285
Halids organic chemistry 547
Hallow-eve customs 394
Halls architecture 725
public hygiene 613
society student life 371
Hallucinations 133
Halogen group chemistry 546
Haloids organic chemistry 547
Halos optics 535
Hamar, Norway history 948.4
travel 914.84
Hamitic language 493
literature 893
Hammers machine tools 621
steam machine tools 621
Hand chiromancy 133
human anatomy 611
language, deaf and dumb 371
railing stair building 694
turning lathes 621
Handbooks of travel **910**
Handicraft mechanic trades **680**
Handkerchief customs 391
Hands, artificial surgery 617
Handwriting 652
Hanging punishments 343
Hangings, textil building 698
Hanover, House of, English hist. 942.07
Hanseatic league history 943
Happiness ethics 171
Harbor defenses fortifications 623
engineering 627
Harbors, hydrografy of surveying 526
Hard labor punishments 343
Hardness of molecules physics 539
Hardware 671
Harelip surgery 617
Harems in Turkey customs 392
Harlots ethics 176
Harmonics analytic geometry 516
modern " 513
music 781
Harmonium musical instruments 786
Harmony music 781
Harmony of colors optics 535
painting 752
gospels Bible 226
Harness-making 685
Harp 787
Harper's Ferry invasion, U. S. history 973.6
Harpsichord 787
Harrison, W: H:, presidency of 973.5
Hartford convention U. S. history 973.5
Harvesting agriculture 633
Hasheesh ethics 178
hygiene 613
Hat-making 687
Hatching artificial, incubation 636
engraving 762
Hate ethics 179
Hats clothing 646
costume 391
Haulage mining engineering 622
Hawaii history 996
travel 919.6
Hawaiian language 499
Hawking sports 799
trade 639
Hay agriculture 633
fever and asthma diseases 616
Hayes, R. B., presidency of 973.8
Hayti history 972.9
travel 917.29
Haze meteorology 551
Hazing student life 371
Head anatomy 611
art anatomy 743
craniology 573
diseases 616
dress customs 391
phrenology 139
surgery 617
Headaches diseases 616
Headstones monuments 718
mortuary design 247
Healing, divine 265
Health affected by soil, hyg. of ground 614
boards of local gov't 352
public health 614
laws hygiene **613**
lift therapeutics 615
of students 371
public **614**
resorts hygiene 613
Healths, drinking of customs 394
ethics 178
Hearing acoustics 534
anatomy 611
asylums for deaf 362

Topics in black-faced type are subdivided in tables.

Hearing diseases 617
ear hospital 362
physiology 612
senses 152
Heart anatomy 611
diseases 616
Heat, animal physiology 612
engines air engines 621
geologic agents 551
meteorology 551
of earth 525
heavenly bodies 523
physics 536
Heathen art ecclesiology 246
missions to 266
philosofy vs the church 261
religions **290**
salvation of 234
Heathenism, non-Christian relig. 299
Heating, buildings 697
domestic econ. 644
houses, sanitary engineer. 628
libraries 022
physics 536
practical chemistry 542
steam engineering 621.1
Heaven theology 237
Hebraic sabbath 263
Hebrew ancient history 933
language 492
literature 892
religion 296
Hebrews epistles 227
modern history 296
Hebrides history 941
Hedges landscape gardening 715
Hedonism ethics 171
Heidelberg catechism 238
Heirs law 347
Heliocentric place 521
Heliometer astronomy 522
Heliostat " 522
Heliotypes 774
Hell 237
Hellebore poisons 615
Hellenic languages, minor 489
Helminthology zoology 595
Helps for laboring classes pol. econ. 331
Hemiptera insects 595
Hemorrhage of the lungs diseases 616
Hemorrhoids " 616
Hemp agriculture 633
manufactures 677
Henri Deux ware pottery 738
Henry 1 English history 942.02
Henry 2 English history 942.03
3 " 942.03
4 " 942.04
5 " 942.04
6 " 942.04
7 " 942.05
8 " 942.05
Hens diseases of 619
poultry 636
Hepaticae botany 588
Hepatology physiology 612
Heptarchy English history 942.01
Heraclitus Greek philosofy 182
Heraldry **929.6**
Herb doctors cures 615
Herbariums 580
Herculaneum description 913.37
Hereditary diseases hygiene 613
genius mental character. 136
succession law 347
pol. sci. 321
Heredity evolution 575
hygiene 613
mental 136
origin of the soul 129
spiritual doctr. theol. 233
Heresies 273
Heretics, persecutions of 272
Hermeneutics Bible 220.6
Hermetic art alchemy 540
Hermits' lives, biografy 920
relig. biografy 922
Hernia diseases 616
Heroism ethics 179
Herpetology 598.1
Herring fisheries 639
Herzegovina history 943.9
Hessian fly economic zoology 591
pests 632
Hexapoda zoology 595
Hibbert lectures religion 290
Hibernation zoology 591
Hicksites friends 289
Hieroglyfics 419
Egyptian 493
High church Anglican church 283
license temperance ethics 178
pressure heating 697
schools 379
treason law 343
Higher education academies 373
colleges 378
Highlands, Scotland, description 914.1
history 941
Hight of atmosfere 523

Topics in black-faced type are subdivided in tables.

Hights, measurement of surveying 526
Highwaymen law 343
Highways, law 347
local government 352
road engineering 625
transportation 386
travel on, protect. of life 614.8
Hildebrand Italian history 945
religious " 270
Himalayan architecture 722
Himalayas physical geografy 551
travel 915.4
Himyaritic language 492
Hindi language 491
Hindostan history 954
Hindu architecture 722
languages 491
law 349
philosofy 181
religion 294
Hinduism religion 294
Hindustan history 954
Hindustani language 491
Histogenesis zoology 591
Histology animals 591
botany 581
human anatomy 611
Historians' lives 928
Historical atlases 911
books Bible 222
charts and tables 902
geografy 911
magazines 905
on special subjects see subject
painting 756
sermons 252
societies 906
on special subjects see subject
theology 209
History **900**
ancient **930**
bibliografy of 016.9
constitutional 342
ecclesiastical **270**
general modern 909
literary, general 809
modern **940-999**
natural **590**
naval, see spec. countries
of civilization 901
philosofy of 901
sacred 220.9
universal 909
of special topics see subject
Histrionics 792
Hives skin diseases 616
Hog domestic animals 636
Hoisting machinery mill work 621
mining engineering 622
Holiday opening library economy 024
Holidays hygiene 613
school, education 371
Holiness doctrinal theology 234
Holland history 949.2
reformation in, relig. hist. 274.9
travel 914.92
Holly system water supply of towns 628
Holstein cattle domes. animals 636
Holtz machine electricity 537
Holy alliance European history 940
communion 265
Ghost knights 929.7
theology 231
grail legends 398
land history, ancient 933
travel 915.6
orders church 265
Roman empire 943
rood ecclesiology 247
scriptures **220**
Spirit theology 231
water ecclesiology 246
Homage feudalism 321
Home customs 392
education self-education 374
vs public 373
life ethics 173
missions 266
relations, customs 392
rule Irish history 941.5
political parties 329
study, societies for 374
woman in 396
Homeopathy therapeutics 615
Homes asylums 725
hygiene 613
residences architecture 728
soldiers' institutions 362
working girls' " 362
laboring classes 331
Homestead bill U. S. law 347
Homicide law 343
Homiletics theology 251
Homilies sermons 252
Homologies biology 574
zoology 591
Homonyms English language 424
French " 444
German " 434
Greek " 484

Topics in black-faced type are subdivided in tables.

Homonyms Italian language 454
Latin " 474
Spanish " 464
Homoptera entomology 595
Honduras, Central America, history 972.8
travel 917.28
Honesty ethics 174
Honey agriculture 638
Honeymoon customs 392
Honor business ethics 174
student 371
Honorary degrees colleges 378
titles heraldry 929.7
Hood, Robin legends 398
Hoods, academic 378
Hooping cough diseases 616
Hoosac tunnel, engineering 622
Hopkinsianism Calvinism 284
Hops agriculture 633
brewing 663
Horn wind instrument 788
organic chemistry 547
Hornd cattle agriculture 636
Horns, drinking customs 394
Horology astronomy 529
watch making 681
Horoscope astrology 133
Horse cars railroad engineering 625
diseases veterinary medicin 619
domestic animal 636
racing amusements 798
ethics 175
railway, engineering 625
sociology 386
useful arts 656
shoeing 682
training, domestic animals 636
veterinary medicin 619
zoology 599
Horsemanship 798
Hortatory theology 243
Horticultural societies 716
agriculture 630
Horticulture 716
fruits and vegetables 634
Hosea Bible 224
Hosiery clothing 646
costumes 391
manufacture 677
Hospice of St Bernard, protec. of life 614.8
Hospital sketches Amer. civil war 973.7
Hospitality ethics 177
Hospitallers knights 929.7
Hospitals 362
and asylums, arch. 725

Host eucharist 265
Hot springs physical geology 551
water heating 697
Hotels 647
architecture 728
hygiene 613
Hothouses agriculture 635
flower culture 716
Hour glasses horology 529
Hours of labor political economy 331
school 371
House building 690
furnishing domestic econ. 645
of commons legis. bodies 328
correction 364
lords legis. bodies 328
refuge 364
representativs legis. bodies 328
painting building 698
sewerage 628
Household art, decoration 747
Housekeeping 640
cooperativ 334
Houses architecture 720
hygiene 613
noble heraldry 929.7
sewerage 628
society student life 371
Housewifery 640
Housing of animals animal hygiene 614
Howitzers ordnance 623
Huckster licenses local gov't 352
Hudson bay territory history 971
river description 917.47
Huguenots church 284
lives 922
persecutions 272
Hulsean lectures apologetics 239
Human ability, doctrinal theology 233
anatomy 611
body and mind 130
body art anatomy 743
depravity doctrinal theology 233
faculties psychology 150
habitations hygiene 613
life, protec. fr. accident 614.8
mind psychology 150
nature dogmatics 233
mental characteristics 136
physiology 612
race ethnology 572
resorts hygiene 613
rights polit. science 323
sacrifice, non-Christ. relig. 290
Humane societies 361

Topics in black-faced type are subdivided in tables.

Humanism		880
Humanitarianism	unitarianism	288
Humanity	ethics	177
Humbugs	delusions	133
Humility	ethics	179
Humming-birds	zoology	598
Humor	American literature	817
	English "	827
	French "	847
	German "	837
	Greek "	887
	Italian "	857
	Latin "	877
	Spanish "	867
Humorists' lives		928
Hungarian language		494
Hungary	history	943.9
	travel	914.39
Hunger and thirst	physiology	612
Huns	history	950
Hunting	business	639
	sport	799
Huronian strata	geology	551
Hurricanes		551
Husband and wife	law	347
Husbandry	**agriculture**	**630**
Husbands, duties of	ethics	173
Hussites	Christian sects	284
Huzwaresh language		491
Hydraulic engine	mech. engin.	621
	physics	532
engineering		627
mining	engineering	622
press	"	621
	physics	532
ram	engineering	621
Hydraulics	science	532
Hydrocarbons	chemistry	547
Hydrocefalus	diseases	616
Hydrodynamics	physics	532
	water engines	621
Hydrofobia	diseases	616
	protectiv inoculation	614
	public health	614
Hydrogen	chemistry	546
Hydrogeology		551
Hydrografy	surveying	526
Hydrokinetics		532
Hydrology		551
Hydromechanics		532
Hydrometer	specific gravity	532
Hydrometry	velocity	532
Hydropathy		615
Hydrostatics		532
Hydrozoa	zoology	593
Hygiene		613
	mental	131
	military	613
	naval	613
	of air and ground public health	614
	animals public health	614
	travel and exploration	613
	prison	613
	public	**614**
	school education	371
	vocal	784
Hygrometry	meteorology	551
Hylozoism	metaphysics	146
Hymn tunes	music	783
Hymnody	"	783
Hymnology	sacred theology	245
Hymns and psalms	hymnology	245
Hyperboreans	Greek mythology	292
Hypnotics	therapeutics	615
Hypnotism	mesmerism	134
Hypochondria	medicin	616
	mental disease	132
Hypotheses	logic	167
Hypothesis, nebular		523
Hypsometry	physical geografy	551
Hysteria	diseases	616
Hysterics	mental derangements	132
Ice age	geology	551
dealers	trades	689
erosion and deposition		551
frost	geologic agents	551
impure	hygiene	613
manufacture		621
meteorology		551
Icebergs	physical geografy	551
Iceland	history	949.1
	travel	914.91
Icelandic language		439
literature		839
Ices	domestic economy	641
Ichnology	paleontology	566
Ichthyology		597
Iconoclasts	heathen art	246
Iconografy		**730**
	Christian	247
Iconology	emblems	246
Idaho, U. S.	history	979.6
	travel	917.96
Idealism	philos. system	141
Idealistic painting		753
Ideas, innate	intuitiv faculty	156
Identity, mistaken	law	347
	personal	126
Idiocy		132

Topics in black-faced type are subdivided in tables.

Idioms	English language	423
	French "	443
	German "	433
	Greek "	483
	Italian "	453
	Latin "	473
	Spanish "	463
Idiosyncrasies, mental		137
Idiot **asylums**		362
schools		371
Idleness	ethics	179
Idolatry	**non-Christ. relig.**	**290**
Igneous rocks	lithology	552
Iguanodon	fossils	568
Illegitimacy	law	347
Illinois, U. S.	history	977.3
	travel	917.73
Illiteracy	public schools	379
	spelling reform	421
Illuminated books		096
manuscripts		091
Illuminati	society	366
Illuminating gases	chem. tech.	665
	public health	614
Illumination, decoration		745
electric		537
	engineering	621.3
house lighting		644
instruments, astronomy		522
sacred ornamentation		247
Illusions		133
Illustrated books	book rarities	096
Illustrations of books	publishing	655
	rare	096
Image worship	**idolatry**	**290**
Images in churches	ecclesiology	246
Imaginaries, calculus of		517
Imaginary books		098
quantities	algebra	512
Imagination in art		701
	science	501
	mental faculties	155
Imbecility	" derangements	132
Imbecils, asylums for		362
education of		371
Immaculate conception of Christ		232
	Mary	230
Immersion	baptism	265
	baptist doctrins	230
	sect	286
Immigrants	colonies	325
Immigration		325
Immoral art and literature	ethics	176
Immortality	doctrinal theology	237
	metaphysics	128
Immortality	natural theology	218
	of animals	591
Impeachment, administration		351
	constitutional law	342
	criminal law	343
Imperial federation, Gt. Br.		354
robes	ceremonies	394
Implements, agricultural		**630**
	domestic	**643**
Importation duty free for colleges		337
	of books, lib. econ.	025
Imports	commerce	382
	duty on	337
Impostors	humbugs	133
Imprecations	ethics	179
Impressionist school of painting		759
Impressment of seamen		359
Imprints, fictitious		014
Imprisonment	punishments	343
Impure ice	hygiene	613
Impurities of well water	hygiene	613
	waterworks	628
Inability, human	doctrinal theology	233
Inaugurations	ceremonies	394
Incandescence, heat		536
Incantations		133
Incarnation	Christology	232
Incas	history	985
Incense	ecclesiology	247
Incest	criminal law	343
Inclined plane, canal engineering		626
	machines	531
Inclines	railroad engineering	625
Income, public	finance	336
tax	"	336
Incubation, artificial		636
Incunabula	book rarities	093
Incurables, hospitals for		362
Indelible inks, chemical technology		667
Indemnity	international law	341
Independence, American	U. S. hist.	973.3
Independency, church polity		262
Independents	sects	285
Index error	practical astronomy	522
expurgatorius	book rarities	098
making, literary methods		029
rerums	" "	029
Indexes, Biblical	concordances	220.2
	see subj. index	
India	antiquities, ancient	913.34
	modern	913.5
	architecture ancient	722
	Mohammedan	723
	botany	581.9
	British history	954

India, Farther history 959
religious history 275.9
geology 555.4
history, ancient 934
modern 954
religious history 274.4
rubber manufactures 678
travel 915.4
Indian archipelago history 991
Bureau U. S. admin. 353
corn agriculture 633
East, languages 491
modern languages 491
legends 398
missions 266
ocean travels 910
philosofers 181
religions 299
Territory, U. S. history 976.6
travel 917.66
wars U. S. hist. 973.2
Indiana, U. S. history 977.2
travel 917.72
Indians, American 970.1
biografy 970.2
education of 371
language 497
literature 897
Indic languages 491
Indicator for steam engines 621.1
Indicators library economy 025
Indictments criminal law 343
Indies, East, Dutch history 991
Indigestion medicin 616
Indigo chemical technology 667
Individual soul, origin of 129
Individuals and the state ethics 172
Indo-European languages, minor 491
literature 891
Germanic languages 491
Indoor amusements 793
Indorsement law 347
Induction electricity 537
magnetism 538
reasoning faculty 153
Inductiv logic 161
sciences 500
Indulgencies Roman church 282
doctrin 230
Industrial arts 600
classes political econ. 331
depressions " " 338
drawing 744
economy 330
education 371
Industrial education useful arts 607
exhibitions 606
history 609
legislation polit. econ. 338
museums 606
sanitation 628
schools 607
statistics political econ. 331
Industries chemical technol. 660
useful arts 600
Industry ethics 179
Inebriate asylums ethics 178
Inebriates temperance 178
Infallibility, divine 231
of the church 262
papal eccles. polity 262
Roman doctrin 230
Roman church 282
schism old catholics 284
Infancy, psychology of 136
Infant baptism 265
education elemen. educ. 372
mortality public health 614
salvation 234
Infanticide customs 392
ethics 173
law 343
Infantry army 356
Infants, care of 618
food for hygiene 613
nursery 649
Infectious diseases pathology 616
pub. health 614
Infernal machines explosivs 662
Infidelity 211
Infinit metaphysics 125
Infinitesimal calculus 517
Infirmaries hospitals 362
Inflammation physiology 612
surgery 617
Inflation paper money 332
Inflection English philology 425
Influenza diseases 616
Infusoria zoology 593
Infusorial earth petrografy 552
Ingenuity inventions 608
Inheritance law 347
of soul metaphysics 129
Inherited disease hygiene 613
mental disability hygiene 613
Injections collectors' manuals 579
medicin 615
Injectors steam engine 621.1
Injured, aid to, protection of life 614.8
Injuries, effects of diseases 616

Topics in black-faced type are subdivided in tables.

Injuries, surgery 617
torts 347
Injurious animals agriculture 632
zoology 591
insects agriculture 632
occupations hygiene 613
plants botany 581
Ink manufacture 667
Inland navigation canals, etc. 656
quarantine public health 614
seas physical geografy 551
Inlay architectural decoration 729
Innate ideas 156
Inns architecture 728
hotels 647
of court English law 340
Inoculation against smallpox 614
protectiv, pub. health 614
Inorganic and organic matter, differ. 577
chemistry 546
Inquisition persecutions by 272
Insane asylum, architecture 725
criminal, law 340
hospitals 362
lives of 920
Insanity law 347
mind and body 132
Inscriptions, comparativ 417
English language 421
Insectivora mammals 599
Insects, fertilization of plants 581
injurious agriculture 632
zoology 591
metamorfosis 591
paleontology 565
zoology 595
Insessores birds 598
Insolvency law 347
Insomnia disease 616
hygiene 613
Inspection of buildings 614.8
explosivs, etc. 614.8
food, etc. pub. health 614.3
Inspiration Bible 220.1
Installation of ministers 262
Instinct animals 591
human 158
Instinctiv knowledge 158
Institute of France 064
Institutes Roman law 349
teachers' 370
Institutions Christian 260
educational 370
feudal 321
monarchic 321
Institutions musical 780
of credit pol. econ. 332
patriarchal 321
republican 321
sociology 360
Instruction education 370
methods of 371
Instrumental errors practical astr. 522
music 780
public worship 264
Instrumentation music 781
Instruments, drawing 744
making 681
for special purposes see subject
Insular faunas, zoology 591.9
floras botany 581.9
Greece ancient history 939
modern " 949.9
Insurance 368
compulsory polit. econ. 331
law 368
societies cooperativ 334
Insurrections internal relations 323
Intaglios sculpture 736
Intarsia and inlay arch. decoration 729
Integral calculus 517
Integrity business ethics 174
Intellect 151
Intellectual development education 370
history 809
influence of ch. 261
relation of ch. 377
life self-education 374
philosofy 150
Intelligence mental science 151
of animals 591
Intemperance ethics 178
Intercommunication, sociology 380
useful arts 650
Interdicts excommunication 262
Interest and usury law 347
calculation of arithmetic 511
political economy 332
securities finance 336
tables business manuals 658
Interference optics 535
Interior decoration 747
department U. S. admin. 353
of earth geology 551
Interlude English drama 822
Intermarriages of families hygiene 613
races ethics 173
Intermediate state theology 236
Interment customs 393
premature 614

Topics in black-faced type are subdivided in tables.

Interment public health 614
Intermittent fever pathology 616
public health 614
Intermolecular forces physics 539
Internal heat geology 551
improvements 620
relations church and state 323
revenue laws finance 336
International association, the 335
coinage 332
congresses 341
on special topics see subject
copyright private law 341
publishing 655
ethics 172
exchange money 332
exchanges lib. econ. 025
exhibitions 606
law 341
standard, money 332
Interoceanic canals engineering 626
sociology 386
Interstate commerce commission 385
law 385
Intervention political science 327
Intolerance, relig., persecutions 272
state ethics 172
Intonation music 781
Intoxicating drinks, traf. in, ethics 178
Intoxication alcoholism 616
ethics 178
Intransigentes, Spanish history 946
Intrenchments military engineering 623
Intuition ethics 171
Intuitionalism, philosofic systems 143
Intuitiv faculty 156
philosofy 143
Inundations meteorology 551
protection engineer. 627
Invalids medicin 616
Invention of printing 655
rhetoric 808
Inventions patents 608
Inventors' lives 926
Invertebrates, paleontology 562
zoology 592
Investiture ecclesiastical polity 262
Investitures, war of, 1074-84 270
Investments 332
Invincible armada Spanish history 946
Invocation of saints and angels 264
Invoices accounts 657
Involution algebra 512
arithmetic 511
geometry 513
Iodin chemistry 546
Ionic Greek dialects 487
order of architecture 729
philosofy 182
Iowa, U. S. history 977.7
travel 917.77
Ipecac emetics 615
Iranian languages 491
Ireland antiquities 913.4
botany 581.9
geology 554.1
history 941.5
religious history 274.1
travel 914.15
zoology 591.9
Iridium chemistry 546
Irish bards lives 928
bulls Eng. humor 827
church, Anglican 283
Roman catholic 282
home rule history 941.5
polit. parties 329
land question, political econ. 333
language 491
music 784
politics 329
Iron age prehist. arch. 571
architecture 721
bearing waters economic geol. 553
bridge engineering 624
building material 691
chemistry 546
clads naval arch. 623
manufacture 672
mask, man in biografy 920
materia medica 615
metallurgy 669
mining 622
ores economic geology 553
pipes waterworks 628
roofs building 695
ships 623
structures arch. con. 721
water pipes sanitary eng. 628
work manufactures 672
Irony rhetoric 808
Irrigation agriculture 631
canal engineering 626
sewage 628
Irritability in plants botany 581
Irvingism catholic apostolic 289
Isaiah Bible 224
Islam Mahometanism 297
Islands, isolated history 997
physical geografy 551

Topics in black-faced type are subdivided in tables.

Isle of Man, England	history	942
Wight, "	"	942
Isles of Shoals	history	974.2
	travel	917.42
Isoeteae	botany	587
Isografy		652
Isomerism	chemistry	541
Isometric projections	geometry	515
Isomorfism	crystallografy	548
Isothermal lines, physical geografy		551
Israelites	history, ancient	933
	modern	296
Issues	therapeutics	615
Isthmian games	Greek antiquities	913.38
Italian	cyclopedias	035
	essays, collected	045
	grand opera	782
	journalism	075
	language	450
	literature	850
	newspapers	075
	painting	759
	periodicals, general	055
	philosofy	195
	republics aristocracies	321
	history	945
	school of Greek philosofy	182
	sculpture, medieval	734
	modern	735
	societies, general	065
Italic languages, minor		479
Italy	ancient history	937
	antiquities	913.4
	botany	581.9
	geology	554.5
	history	945
	reformation in	274.5
	religious history	274.5
	travel	914.5
	zoology	591.9
Ivory-carving		736
supply		679
Jackson, Andrew, presidency of		973.5
Jacobins	French history	944
	political clubs	363
Jacobite church		281
Jacobites	Eng. pol. party	329
Jacotot's system of education		371
Jacquerie	French history	944
Jade	mineralogy	549
Jails	prisons	365
Jaina architecture		722
Jainas	Hindu sect	294
Jamaica	history	972.9
James, St	epistle	227
James 1	English history	942.06
2	" "	942.06
Jansenism	heresy	273
Japan	botany	581.9
	chemical technology	667
	geology	555.2
	history	952
	religious history	275.2
	travel	915.2
Japanese architecture		722
language		495
Japanning	varnishing	698
Jasher, book of		229
Java	history	992
	travel	919.2
Jealousy	ethics	179
Jeanne d'Arc	French history	944
Jeddo	travel	915.2
Jefferson, T:, presidency of		973.4
Jehovah	doctrinal theology	231
Jelly fish	zoology	593
Jeremiah	Bible	224
Jeremy, Epistle of	Bible	229
Jersey City, N. J.	history	974.9
	travel	917.49
Island of	history	942
	travel	914.2
Jerusalem	history, ancient	933
	St John of knights	929.7
	travel	915.6
Jesters	customs	394
Jests	English literature	827
Jesuit missions		266
Jesuits	religious orders	271
Jesus	Christology	232
	society of religious orders	271
Jet	economic geology	553
Jetties	engineering	627
Jewelry	customs	391
	manufacture	671
	toilet	646
Jewish architecture		722
	calendar	529
	history	933
	language	492
	literature	892
	philosofy	181
	religion	296
Jews, disabilities of		296
	history, ancient	933
	modern	296
	restoration of	296
Job	Bible	223
Jockies	horse racing	798
Joel	Bible	224

Topics in black-faced type are subdivided in tables.

John, King	English history	942.03
John, St	epistles	227
	gospel	226
	knights of	929.7
	Revelation of	228
Johnson, A., presidency of		973.8
Joinery, ornamental	carpentry	694
Joint stock companies	law	347
Jokes	English humor	827
Jonah	Bible	224
Joshua	Bible	222
Journalism		**070**
Journalists' lives		920
Journeys around the world		910
Jubilee singers of Fisk university		784
Judaism		296
vs Christian church		239
Jude	epistle Bible	227
Judea	ancient history	933
	antiquities	913.33
Judges, book of	Bible	222
law courts		347
lives		923
Judgment, mental faculties		153
private, ecclesiastical polity		262
theology		236
Judicial chemistry	law	340
combat	"	340
control	administration	351
ordeal	law	340
Judiciary department	U. S. gov't	353
electiv	law	347
Judith	Bible	229
Juggernaut		299
Jugglery		791
Julian calendar		529
Jupiter	astronomy	523
	mythology	292
Jurassic age	geology	551
Jurisdiction	civil trials	347
Jurisprudence		**340**
	medical	340
Jury trial	law	340
Justice	ethics	179
Justices of the peace	law	347
Justification	doctrinal theology	234
Jute	agriculture	633
	manufactures	677
Juvenil books,	reading and aids	028
	see subject	
criminals	law	343
literature		028
mortality	public health	614
reformatories		364
Kabala	Jewish philosofy	181
Kafirs	history	968
Kaleidoscope	optics	535
Kalmucks	description	915
	history	950
Kamtchatka	history	957
Kangaroo	zoology	599
Kansas, U. S.	history	978.1
	travel	917.81
struggle in	U. S. hist.	973.6
Kant	German philosophy	193
Keltic antiquities		913.36
Druids religion		299
languages		491
Kelts ancient history		936
Kempis, Thomas à		242
Kentucky, U. S.	history	976.9
	travel	917.69
Keramics	art	738
	chemical technology	666
Kerosene	" "	665
	inspection, protect. of life	614.8
Keys and locks	manufactures	683
mathematical	students' honor	371
Khiva	travel	915.8
Kidnapping	criminal law	343
Kindergarten	education	372
Kindness	ethics	177
Kinematics	physics	531
Kinetic theory of gases		533
Kinetics	physics	531
King Philip's war	history	974
Kings bench		347
book of, Bible		222
daughters		267
lives		923
political science		321
Kinship	genealogy	929
Kissing	customs	392
Kitchen economy		**640**
gardening		635
middens	archeology	571
Kitchens	architecture	728
Kleptomania	mental disease	132
Knightage	heraldry	929.7
Knights hospitallers		929.7
of labor,	labor organizations	331
Malta		929.7
Pythias	secret societies	366
Rhodes		929.7
St John		929.7
Templars		929.7
Knitting	domestic economy	646
Knives	cutlery	672
	stone	571
Know-nothing party	U. S. politics	329

Topics in black-faced type are subdivided in tables.

Knowledge, theory of		121
unity of		121
Koordistan	travel	915.6
Koran	**Mahometanism**	297
Krishna	Buddhism	294
Kroatian language		491
Kuklux Klan, political associations		363
Kurdish language		491
Kurdistan	travels	915.6
Kurile islands	travel	915.2
Labels in museums		579
Labor	laws	338
	obstetrics	618
	of children	331
	party political parties	329
	political economy	331
	savers, literary	029
	statistics political econ.	331
Laboratories	architecture	727
	biologic	590
	school equipment	371
	state adulterations	614.3
Laboratory, chemistry		542
	work, methods of instruc.	371
Laborers' cottages	architecture	728
	occupation hygiene	613
Laboring classes	political economy	331
Labrador	history	971
Labyrinths	prehistoric archeology	571
Lace-making		677
	art needlework	746
Laconics	English literature	828
Lacquer work,	artistic furniture	749
	manufactures	667
Lacrosse	games	797
Lactometer		614.3
Ladies of the Sacred Heart		271
Lake drainage		627
	dwellings prehistoric arch.	571
	states, U. S. **history**	977
	travel	917.7
	surveys geodesy	526
Lakes, artificial, landscape gardening		714
	physical geografy	551
	water supply	628
Lamaism		294
Lambertypes		773
Lamellibranchiata	**zoology**	594
Lamentations	Bible	224
Lampoons	English satire	827
Lampreys	fishes	597
Lamps	chemical technology	665
Lancaster and York	Eng. hist.	942.04
	system education	371
Land claims	finance	336
Land grants	finance	336
	to public schools	379
	law	347
	laws political economy	333
	political economy	333
	public political economy	333
	slides geology	551
	surveying	526
	tenure law	347
	titles, insurance of	368
Landed gentry	English heraldry	929.7
Landlord and tenant	law	347
	polit. economy	333
Landscape	drawing	741
	gardening	**710**
	painting	758
	photografs	779
	sketching	741
Landtag	German central gov't	354
	local "	352
Language		**400**
	English, origin of	422
	lessons elemen. educ.	372
	of flowers	716
	origin of	401
	sign deaf and dumb	371
	hieroglyfics	419
	universal	408
Lantern, astronomic		523
	magic optics	535
La Plata	history	982
	travel	918.2
Larceny	law	343
Lard	adulterations	614.3
	chemical technology	664
Larva	embryology	591
Laryngoscope		616
Laryngotomy	surgery	617
Larynx	acoustics	534
	anatomy	611
Late hours	hygiene	613.7
Lath	plastering	693
Lathe work	mechanical **eng.**	621
Latin antiquities		913 37
	biografy	920
	church	282
	classics	**870**
	epic poetry	873
	erotica	874
	geografy	912
	hymns **hymnology**	245
	inscriptions	471
	language	**470**
	literature	**870**
	lyric poetry	874

Topics in black-faced type are subdivided in tables.

Latin, medieval and modern 479
mythology 292
union 332
Latitude at sea navigation 527
determinations geodesy 526
finding 525
Latter-day saints Mormonism 298
Laughing gas anesthetics 615
Laughter emotions 157
hysteria 616
Laundry domestic economy 648
Lava lithology 552
Lavatories school furniture 371
Law 340
addresses 347
and miracles theology 231
order leagues temperance 178
moral ethics 170
of nations 341
nature 340
special topics see subject
parish ecclesiastical polity 262
reports American 345
English 346
schools 340
architecture 727
Sunday 263
Lawn-tennis 796
Lawns 712
Laws, building building 692
protection of life 614.8
conflict of 341
medical license 614
Laws of language Eng. etymology 422
Lawyers ethics 174
lives 923
Lay ministry ecclesiastical polity 262
Laying out of towns and streets, eng. 625
public health 614
Lazarettos contagious diseases 614
hospitals 362
Lazarists monastic orders 271
Lead, black economic geology 553
chemistry 546
drugs 615
manufacture 671
metallurgy 669
mining 622
ores economic geology 553
poisoning toxicology 615
roofs building 695
water pipes 628
Leading cases, American law 345
English " 346
on special subjects see subject
Leading glazing 698
Leagues, law and order temperance 178
Learned societies 060
Leases law 347
Leather manufacture 675
stampt building 698
Leaves, shapes of botany 581
Lecompton constitution Kan. hist. 978.1
U. S. " 973.6
Lectern sacred furniture 247
Lectionary Anglican ritual 264
Lectures English oratory 825
for laboring classes 331
halls for architecture 725
instruction methods 371
library 021
popular education 374
on special topic see subject
Leeches medicin 616
zoology 595
Legacies law 347
taxation of 336
Legal anecdotes 340
bibliografy 016.3
capacity of persons law 347
education 340
ethics 174
hermeneutics 347
maxims 340
status of woman 396
tender paper money 332
Legends folk-lore 398
monastic 271
mythology 291
of the Madonna 225
Legerdemain 791
Legislation, American statutes 345
British " 346
comparativ 340
industrial 338
library, founding of libs. 021
sumptuary, social ethics 177
on special topic see subject
Legislativ bodies and annals 328
Legitimacy family law 347
Legs, artificial surgery 617
Leinster, Ireland history 941.8
travel 914.18
Lemons agriculture 634
Lemur zoology 599
Lenses optics 535
Lent church fast 264
Lenten mass music 783
Lepidoptera zoology 595
Leprosy pathology 616

Topics in black-faced type are subdivided in tables.

Leprosy public health 614
Letter-writing 808
Lettering, ornamental 745
Letters American literature 816
English alfabet 421
literature 826
French " 846
German " 836
Greek " 886
Italian " 856
Latin " 876
patent patents 608
Spanish literature 866
on special topic see subject
Lettic languages 491
Lettish " 491
Levant history 956
travel 915.6
Levees canal engineering 627
Leveling geodesy 526
Lever mechanics 531
Leviticus Bible 222
Lex mercatoria law 347
Lexicografy, comparativ 413
English language 423
Lexicons see language
on special topics see subject
Libel law 347
Liberal Christianity rationalism 211
unitarianism 288
Liberality ethics 179
Liberty, civil law 374
political science 323
ecclesiastical polity 262
metaphysics 123
of conscience eccles. pol. 262
press 323
will mental faculties 159
personal doctrin 233
law 347
pol. science 323
salvation 234
religious church 261
state ethics 172
Librarian, duties, etc. 023
Librarians' lives 920
Libraries, architecture 727
lib. economy 022
for work'g class. pol. econ. 331
lib. econ. 027
free public 027
history 027
imaginary book rarities 098
parochial library econ. 027
parish work 257
Libraries, regulations for readers 024
school education 371
library economy 027
scope and founding 021
special library economy 026
polygrafy 080
Sunday opening, lib. econ. 024
theology 263
Libraries, in sense of series, see subject
Library administration 025
buildings 022
architecture 727
catalogs, author 018
dictionary 019
subject 017
economy 020
gov't and service 023
legislation lib. econ. 021
Librettos opera 782
License, local government 352
of dentists pub. health 614
pharmacists " " 614
physicians " " 614
taxation 336
temperance 178
Licenses, maritime commerce 387
Lichens botany 589
Lieder vocal music 784
Lien laws 347
Life, biology 570
boats protection of life 614.8
ship-building 699
conditions of biology 577
contingencies, probabilities 519
elixir of alchemy 540
ethics 179
future doctrinal theology 237
natural " 218
insurance 368
probabilities 519
metaphysics 111
origin of biology 576
preservers protection of life 614.8
properties of living matter 577
protection of, from accidents 614.8
saving service, protection of life 614.8
transportation 656
tables life insurance 368
Light and air hygiene 613
effect on plants 581
electric engineering 621.3
physics 537
houses engineering 627
navigation 656
optics 535

Light houses protection of life 614.8
infantry military science 356
of earth 525
moon, stars and sun 523
optics 535
polarization of 535
therapeutics 615
zodiacal astronomy 523
Lighting, electric engineering 621.3
physics 537
library buildings 022
Lightning and lightning rods 537
shock surgery 617
Lights and fuel domestic economy 644
shades drawing 742
perspectiv 515
chemical technology 665
ecclesiology 246
Lignite economic geology 553
Lima history 985
travel 918.5
Limbo eschatology 236
Limbs artificial 617
Limbus infantum eschatology 236
patrum " 236
Lime building 691
Limes building material 691
chemical technology 666
economic geology 553
Limestone economic geology 553
lithology 552
Lincoln, A., presidency of 973.7
Lincrusta paper-hanging 698
Line engraving 765
Linear perspectiv 515
Linen manufacture 677
Lines analytic geometry 516
descriptiv " 515
geometry 513
Linguistics 400
Linguists' lives 924
Linnaean system botany 580
zoology 590
Lions zoology 599
Liquefaction heat 536
Liquid wastes, indus. sanitation 628
Liquids physics 532
Liquor laws temperance 178
Liquors manufacture 663
taxation 336
traffic in, temperance 178
use of " 178
Lisbon history 946.9
travel 914.69
Litanies sacred music 783
Litany church service 264
Literary addresses 804
on special topics see topic
clubs self-education 374
criticism, art of 801
with thing criticized see subject
esthetics 801
frauds and forgeries 809
history 809
institutions colleges 378
men's lives 928
methods library economy 029
miscellanies, English 828
occupations hygiene 613
periodicals 805
property law 347
publishing 655
societies student life 371
style rhetoric 808
Literature 800
bibliografy of 016.8
biografy of 928
classical, Greek & Latin 880
immoral ethics 176
lib. econ. 025
medieval 800
modern 800
woman delineated in 396
Lithografy engraving 763
Lithology geology 552
Lithuanian language 491
Little sisters of the poor 271
Liturgies 264
Live stock agriculture 636
Liverworts botany 588
Lives Amer. Indians 970.2
biografy 920
gipsies 397
slaves 326
Living, cost of 647
matter, properties of 577
Livonian language 491
Lizards reptils 598.1
Loadstone economic geology 553
magnetism 538
Loan institutes polit. economy 332
systems library " 025
Loango, Africa travel 916.7
Loans finance 336
Lobsters zoology 595
Local government administration 352
histories, see name of place
option 178
taxes finance 336
Location of library buildings 022

Topics in black-faced type are subdivided in tables.

Locke, J: sensational philosofy 145
Locks and keys manufacture 683
canal engineering 626
Locomotion animal comp. anat. 591
communication 656
physiology 612
Locomotivs mech. engineering 621.1
railroading 656
Locusts economic zoology 591
pests 632
Lodestone economic geology 553
magnetism 538
Lodging houses for poor 361
Log book seamanship 656
Logarithms 510
Logging timber trade 674
Logic **160**
Logical catalogs 017
machines 164
Logos Christology 232
Lollards, religious sect 289
Lombard street banking 332
London, England history 942.1
travel 914.21
Longevity life insurance 368
physiology 612
probabilities 519
statistics 312
Longitude at sea navigation 527
determinations geodesy 526
methods of finding 525
theoretic astronomy 521
Loo-choo islands travel 915.2
Loom, power manufactures 677
Lord chancellors' lives 923
Lord's day 263
prayer 226
supper sacraments 265
Lost books book rarities 098
library economy 024
Lotteries ethics 174
history of 174
Lotus Egypt. religion 299
Louisiana, U. S. history 976.3
travel 917.63
Louvre gallery, art galleries 708
Love ethics 177
feasts pub. worship 264
of God theology 231
Low countries history 949.2
Dutch language 439
German literature 839
Latin language 479
literature 879
Lower empire Byzantine 949.5
Lower Silurian group geology 551
Loyalty state ethics 172
Lubrication mechan. engineering 621
Lucernariae zoology 593
Lucretius ancient philosofy 187
Luke gospel 226
Lumber building material 691
manufacture 674
Lunacy law 347
mind and body 132
Lunar caustic, therapeutics 615
methods of finding longitude 525
tables 523
Lunatic asylums 362
Lungs, admin. of med. by 615
anatomy 611
diseases of 616
functions of physiol. 612
Luray caverns, description 551
Lusitania history 946.9
Lute stringd instruments 787
Lutheran church 284
confession creeds 238
Lutherans' lives 922
Luxury hygiene 613
Lycantrophy insanity 132
Lyceum Greek philosofy 185
Lyceums self-education 374
Lycopodinae botany 587
Lying ethics 177
Lying-in hospitals 362
Lynch law 343
Lyric poetry, Greek 884
Latin 874
Mabinogeon Welsh literature 891
Macaronics English literature 827
Maccabees Bible 229
Machine engraving 768
guns military engineering 623
Machinery **engineering** **621**
in industry pol. econ. 338
Machines hydraulic 532
mechanics 531
Mackerel fishery 639
Madagascar geology 556.9
history 969
religious history 276.9
travel 916.9
Madeira islands history 946
Madhouses 362
Madness mind and body 132
Madonna in art 709
mariolatry 230
Madrid travel 914.6
Madrigal music 784

Topics in black-faced type are subdivided in tables.

Magazine guns see Rifles
Magazines **general periodicals** **050**
literary 805
special subject see subject
Magellan straits travels 918.2
Magic lanterns, optics 535
natural 133
Magicians lives 920
Magistrates law 347
Magna charta constitutional history 342
English history 942.03
Magnesium chemistry 546
Magnetic telegraf 654
Magnetism 538
animal, mind and body 134
terrestrial 538
Magneto-dynamos electricity 537
Magnets properties of 538
Magnifiers optics 535
Magyar language 494
Magyars history 943.9
Mahabharata Sanskrit literature 891
Mahometanism 297
Mails post-office 383
Maine, U. S. history 974.1
travel 917.41
Maine law temperance 178
Maize agriculture 633
Majolica pottery 738
Malachi Bible 224
Malacology 594
Malacostraca zoology 595
Malaria hygiene 613
marshes 614
Malarial fever, diseases 616
public health 614
Malarious soils " " 614
Malay archipelago history 991
travel 919.1
Polynesian languages 499
literatures 899
Malaysia geology 559.1
history 991
religious history 279.1
travel 919.1
Malformations, animal comp. anat. 591
human, deformities 617
nat. hist. of man 573
vegetable botany 581
Malleability molecular physics 539
Malpractice law 343
public health 614
Malt liquors brewing 663
Malta history 945
knights of 929.7
Malthusianism population 312
Mamelukes Egyptian history 962
Mammalia paleontology 569
zoology 599
Mammoth paleontology 569
cave physical geografy 551
zoology 599
Man, antiquity of prehist. arch. 571
nat. hist. of man 573
doctrinal theology 233
ethnografy 572
in the iron mask biografy 920
Isle of history 942
mammals 599
mental characteristics 136
faculties **150**
natural history of 573
origin of evolution 575
nat. hist. of man 573
natural theology 213
physiology of 612
Management, library economy 025
of state properties 336
Manasses, prayer of Apocryfa 229
Mandolin stringd instruments 787
Manganese, chemistry 546
ores economic geology 553
Mania, mind and body 132
Maniacs lives 920
Manicheism heresies 273
Manipulation practical chemistry 542
Manitoba, Canada history 971
Manly exercises outdoor sports 796
Manners and customs, general **390**
for local see special countries
Manometer pneumatics 533
Manor houses, architecture 728
Manors political economy 333
Mansard roofs, arch. construction 721
design 729
Mansions architecture 728
Manslaughter law 343
Mantels fixt furniture 729
Manual training education 371
Manuals, business commerce 658
collectors' biology 579
legislativ U. S. 328
special see subject
Manufactories architecture 725
Manufacture of explosivs protection of life 614.8
Manufactures **670**
chem. technology **660**
production 338
public health 614

Topics in black-faced type are subdivided in tables.

Manufacturing works engineering 621
Manures agriculture 631
sewage 628
Manuscript catalogs library econ. 025
Manuscripts book rarities 091
copyright 655
diplomatics 417
Latin diplomatics 471
Manx language 491
Map cases library admin. 025
drawing in schools 910
projection, geodesy 526
Maps 912
astronomic, general 524
moon, planets, star, or sun 523
school apparatus 371
Marathi language 491
Marble economic geology 553
inscriptions 417
Greek 481
Latin 471
lithology 552
sculpture 731
working 693
Marian persecutions in England 272
Marine animals 591.9
architecture naval eng. 623
corps navy 359
engines 621.1
insurance 368
law 347
plants 581.9
Mariners' compass magnetism 538
occupation hygiene 613
Mariolatry worship of Virgin 230
Marionnets amusements 791
Maritime affairs naval science 359
discovery 910
law 347
quarantine public health 614
surveying 526
Mark gospel 226
Market gardening agriculture 635
Markets architecture 725
domestic economy 641
Marking system education 371
Marks political science 321
Marl lithology 552
Maronites sects 281
Maroons history of Jamaica 972.9
Mar-prelate controversy puritans 285
Anglican ch. 283
Marque and reprisals intern. law 341
Marquetry artistic furniture 749
Marquetry cabinet work 684
Marriage customs 392
ethics 173
law 347
registers genealogy 929
sacrament 265
Mars planets 523
Mars' seasons astronomy 525
Marshes, hygiene of public health 614
physical geology 551
Marsupialia mammals 599
Martial law 344
Martyrs persecutions 272
Mary, blessed virgin, Rom. cath. doc. 230
queen of Eng. Eng. history 942.05
Maryland, U. S. history 975.2
travel 917.52
Masks dramatic music 782
Mason and Dixon's line slavery 326
Masonry, anti secret societies 366
building 693
Masons secret societies 366
Masquerades amusements 793
social customs 394
Mass Rom. catholic ritual 264
law of gravity 531
of earth 525
recitative music 784
sacred " 783
Massachusetts, U. S. history 974.4
travel 917.44
Massage therapeutics 615
Mastmaking ship-building 699
Master and servant family ethics 173
law 347
Mastersingers German poetry 831
Mastodon paleontology 569
Matches chemical technology 662
Materia medica 615
Materialism philosofy 146
Materials building 691
for clothing hygiene 613
library bookbinding 025
buildings 022
strength of, engin. 620
molec. physics 539
Maternity ethics 173
hospitals 362
medicin 618
Mathematical drawing 744
instruments 510
probabilities 519
tables 510
of music 781
tides 525

Mathematicians' lives 925
Mathematics **510**
Matins church ritual 264
Matriculation schools 371
Matter metaphysics 117
physics 530
properties of living 577
Matthew gospel 226
Mauritius geology 556.9
history 969
religious history 276.9
travel 916.9
Mausoleums 718
Maxima and minima algebra 512
calculus 517
geometry 513
Maxims English literature 828
law 340
on special topic see subject
Maya architecture 722
language 497
Mayor local gov't 352
Mazurka piano music 786
Meals domestic economy 642
Mean time solar and sidereal day 529
value probabilities 519
Measles diseases 616
public health 614
Measurements, electric 537
practical chemistry 542
Measures and weights, meteorology 389
Measuring instruments " 389
physics 530
of angles, crystallografy 548
heat 536
light optics 535
surveying 526
Meat, inspection of, pub. health 614.3
preservation of chem. tech. 664
supply domestic economy 641
production 338
Meats as food hygiene 613
Mechanic arts and trades **680**
Mechanical drawing 744
engineering **621**
equivalent heat 536
musical instruments 789
remedies therapeu. 615
trades **680**
training 607
in schools 371
Mechanics physics 531
Mechanics' institutes exhibitions 606
liens law 347
lives 926
Mechanism engineering **621**
Medallions numismatics 737
Medals carving 736
numismatics 737
Media ancient history 935
Medical analysis chemistry 543
biografy 926
botany econ. botany 581
materia medica 615
diagnosis 616
education 610
state control of 614
electricity, physics 537
therapeutics 615
ethics 174
jurisprudence 340
museums 610
police 614
practice pathology 616
state control of 614
schools 610
societies 610
testimony 614
zoology 591
Medicin **610**
chemical analysis 543
economic botany 581
zoology 591
preventiv, pub. health **614**
state control of, pub. health 614
veterinary 619
Medicinal use of stimulants temperance 178
Medicins therapeutics 615
patent public health 614.3
Medico-legal relations 614
Medieval architecture 723
art, history of 709
customs and costumes **390**
designs 745
Europe history **940**
Latin 479
philosofy 189
science 502
sculpture 734
Meditations theology 242
Mediterranean sea physical geology 551
travel 910
Mediums spiritualism 133
Medo-Persia ancient history 935
antiquities 913.35
Medusae zoology 593
Meekness ethics 179
Meerschaum econ. geol. 553
Megaric philosofy 183
Megatherium paleontology 569

Topics in black-faced type are subdivided in tables.

Melancholy mind and body 132
Melanesia history 993
travel 919.3
Melbourne history 994
travel 919.4
Meliorism philosofic systems 149
Melodeon music 786
Melody theory of music 781
Memoirs **biografy** 920
scientific societies 506
Memorial day U. S. civil war 973.7
sermons **biografy** 920
sermons 252
Memory in animals zoology 591
mental derangements 132
faculties 154
Menageries amusements 791
zoology 590
Meningitis diseases 616
Mennonite church 289
Mensuration arithmetic 511
surveying 526
Mental capacity of women, education 376
mind and body 136
characteristics 136
derangements medicin 616
mind and body 132
development **education** 370
disability, inherited hygiene 613
faculties 150
of animals zoology 591
healing 615
heredity mind and body 136
hygiene 131
overwork mental hygiene 131
philosofy 150
photografs 139
physiology mind and body 131
science 150
Mercantil law 347
libraries 027
manuals 658
rules arithmetic 511
theory and practice 658
Merchant service, ocean trans., sociol. 387
Merchants business ethics 174
lives 923
Mercury chemistry 546
materia medica 615
metallic poisons 615
metallurgy 669
ores economic geol. 553
planets 523
Mercy, sisters of 271
Meridian line astronomy 522
Meridional instruments 522
Merino sheep domestic animals 636
Mermaids folk-lore 398
Mesmerism mind and body 134
Messenger service 385
Messiah doctrinal theology 232
Metafors rhetoric 808
Metallic ores econ. geol. 553
poisons therapeu. 615
tractors cures 615
Metallurgy 669
Metals building materials 691
coinage 332
history 669
inorganic chemistry 546
manufactures 671
mining 622
organic chemistry 547
transmutation of 540
Metamorfic rocks geology 552
Metamorfism physical geology 551
Metamorfosis botany 581
of insects 595
zoology 591
Metaphysics 110
Metempsychosis metaphysics 129
Meteoric showers astronomy 523
Meteorites lithology 552
Meteorology 551
natural history of man 573
Meteors and comets astronomy 523
Meter weights and measures 389
Meters Eng. prosody 426
water 628
Method of exhaustions calculus 517
least squares orbits 521
probabil. 519
Methodist church 287
episcopal church 287
Methodists' lives 922
Methodology metaphysics 112
Methods, literary lib. econ. 029
of education 371
statistics 311
in special topics see subject
Metric system intern'l measures 389
Metrology 389
Mexican Indians 970.1
war U. S. history 973.6
Mexico antiquities 913.7
botany 581.9
geology 557.2
gulf of phys. geol. 551
history 972
religious history 277.2

Mexico travel 917.2
Mezzotint 766
Mica economic geology 553
Micah Bible 224
Michigan, U. S. history 977.4
travel 917.74
Microbes botany 589
etiology 616
Microfone 537
Micrometers astronomy 522
Micronesia history 996
Micro-organisms theory of disease 616
Microphotografy microscopy 578
photografy 778
Microscopes 578
optics 535
Microscopic analysis of water 613
life invertebrates 592
Microscopy 578
animal histology 591
lithology 552
study of disease 616
vegetable histology 581
Microtome microscopy 578
Middens and privies scavenging 628
kitchen prehist. arch. 571
Middle ages European history 940
high German language 437
poetry 831
states, U. S. history 974
Middlesex, England " 942.1
Midlands, " " 942
Midwifery obstetrics 618
Migrations of animals 591
men ethnology 572
Mildew botany 589
Military academies, e. g. West Point 355
and naval arts 355
antiquities 355
biografy 923
engineering 623
history 355
Amer. civil war 973.7
of special countries see country
law 355
organization in schools 371
pensions 351
religious orders 929.7
science 355
signals 623
Militia 355
Milk analysis adulterations 614.3
chemistry 543
and its products adult. 614.3
condenst 637
Milk dairy 637
Mill dams hydraulic eng. 627
water engines 621
work engineering 621
Millennium theology 236
Millerites 289
Millinery clothing 646
Milling tools 621
Mills and manf. works engineer'g 621
architecture 725
cotton, woolen, etc. 677
paper manufactures 676
Milwaukee, Wis. history 977.5
travel 917.75
Mimetic arts 792
Mimicry 177
Mind 150
and body 130
cure therapeutics 615
in animals zoology 591
readers' lives 920
reading mesmerism 134
Mineral cements chem. technology 666
economic geology 553
irritant poisons, therapeu. 615
oils chemical technology 665
economic geology 553
springs economic geol. 553
surveys 622
veins ore deposits 553
waters adulterations 614.3
artifi. chem. tech. 663
economic geology 553
hygiene 613
therapeutics 615
Mineralogy 549
of the Bible 220.8
Minerals rock determination 552
Mines military engineering 623
mining 622
ownership, political economy 333
Miniatures painting 757
Minima and maxima algebra 512
calculus 517
geometry 513
Mining companies mining eng. 622
dangers and accidents, eng. 622
engineering 522
laws 622
products political economy 338
Minister eccles. polity 262
Ministers foreign government 354
lives 922
U. S. government 353
Ministration to sick and dead 265

Ministries of state architecture 725
Ministry ecclesiastical polity 262
  lay 262
  parish work 258
  religion 250
Minnesingers German poetry 831
Minnesota, U. S. history 977.6
  travel 917.76
Minority representation polit. sci. 324
Minors law 347
Minstrelsy English poetry 821
  negro vocal music 784
Mints political economy 332
Minute books curiosa 099
Miocene period geology 551
Miracle plays Eng. literature 822
  general history 809
  religion 244
  theater 792
Miracles doctrins 231
  New Testament 226
  post-biblical 231
Mirage optics 535
Mirrors " 535
Miscegenation ethics 173
  hygiene 613
  law 347
Miscellanies see subjects
Miscellany American literature 818
  English " 828
  French " 848
  German " 838
  Greek " 888
  Italian " 858
  Latin " 878
  Spanish " 868
Misers' lives 920
Mishna Judaism 296
Missals illuminated 096
  Rom. catholic ritual 264
Missing link origin of species 575
Missionaries' lives 922
Missionary schools 377
Missions 266
  music for sacred music 783
  parish 269
Mississippi, U. S. history 976.2
  travel 917.62
  bubble loan institutes 332
  river description 917.7
  valley " 917.7
Missouri, U. S. history 977.8
  travel 917.78
  river, description 917.78
Mites zoology 595

Mithraic worship 295
Miter sacred ornaments 247
Mixt drinks adulterations 614.3
  chemical tech. 663
Mnemonic derangements 132
Mnemonics 154
Moabite stone 492
Modeling sculpture 731
Models art study 707
Modern architecture 724
  costumes and customs 390
  Greek language 489
  literature 889
  history 940–999
  Persian languages 491
  philosofers 190
  lives 921
  sculpture 735
Modesty ethics 179
Moguls Indian history 954
Mohammedan architecture 723
Mohammedanism 297
Moisture hygiene 613
  physical geografy 551
  soil public health 614
  therapeutics 615
Molasses chemical technology 664
Mold fungi 589
Moldavia, Turkey history 949.8
  travel 914.98
Moleculat physics 539
Molecules theoretic chemistry 541
Moles zoology 599
Molinists heresies 273
  persecution 272
Mollusca paleontology 564
  zoology 594
Molluscoidea 594
Molly Maguires labor organizations 331
Moluccas history 991
Moments statics 531
Monachism theology 271
Monaco history 944
Monarch foreign governments 354
Monarchic institutions 321
Monarchs' lives 923
Monarchy 321
Monasteries architecture 726
  theology 271
Monastic orders " 271
  schools education 377
Monetary congresses 332
Money 332
  making commerce 658
  ethics 174

Money paper 332
Mongolian languages 495
Mongols history 950
Monism philosofy 147
Monitorial system of education 371
Monitors, iron clads naval eng. 623
school government 371
Monkeys zoology 599
Monks and nuns relig. orders 271
Monocotyledonae botany 584
Monogamy family ethics 173
hygiene 613
Monograms design 745
Monoliths prehistoric arch. 571
Monomania insanity 132
Monopolies 338
state, e. g. tobacco in France 336
Monotremata mammals 599
Monroe, J., presidency of 973.5
Monroe doctrin foreign relations 327
Monsoons meteorology 551
Monstrosities, animal 591
human dwarfs, etc. 573
vegetable 581
Mont Blanc travels 914.4
Mont Cenis tunnel 622
Montana, U. S. history 978.6
travel 917.86
Montanism heresies 273
Montenegro history 949.7
travel 914.97
Months, names of 529
Montreal history 971
travel 917.1
Monts de piété loan institutes 332
Monuments **antiquities** **913**
cemeteries 718
prehistoric 571
Moon astronomy 523
charts 523
effect on tides 525
maps observations 524
Moorish architecture 723
Moot courts 340
Moraines glaciers 551
Moral agency, free, doctrinal theology 233
education 377
expectations probabilities 519
government in schools 371
heredity doctrin 233
metaphysics 129
philosofy **170**
science **ethics** **170**
sentiment " 171
systems 171
Morals and church 261
history of ethics 170
laboring classes polit. econ. 331
of the press ethics 179
war " 172
Moravia history 943.7
Moravian brethren sects 284
language 491
Morbid anatomy 616
Morea history 949.5
Moresnet " 949.3
Morfology animals 591
comparativ philology 415
English language 425
plants 581
Morganatic marriage ethics 173
customs 392
Mormonism 298
Morning prayer church ritual 264
family devotions 249
personal religion 248
Morocco geology 556.4
history 964
religious history 276.4
travel 916.4
Morphin habit ethics 178
Mortality, law of insurance 368
statistics 312
public health 614
Mortar building material 691
Mortars chemical technology 666
ordnance 623
prehistoric archeol. 571
Mortgages law 347
Mortification surgery 617
Mortmain law 347
Mortuary buildings, architecture 726
customs 393
designs sacred ornaments 247
Mosaic law Old Testament 222
work 747
Mosaics arch. decoration 729
floors and flooring 721
religious ornaments 247
Moslems 297
Mosques architecture 726
Mosquito shore, Cent. Amer., history 972.8
Mosquitos zoology 595
Mosses botany 588
sea algae 589
Motet sacred music 783
Mother ethics 173
Goose folk-lore 398
maternity 618
Moths domestic economy 648

Topics in black-faced type are subdivided in tables.

Moths economic zoology 591
zoology 595
Motion, beginnings of origin of life 576
in medicin massage 615
laws of astronomy 521
metaphysics 116
of earth astronomy 525
liquids hydraulics 532
plants botany 581
solar system 523
Mottos English literature 828
Mould fungi 589
Mound builders prehistoric arch. 571
U. S. history 973.1
Mounds prehistoric arch. 571
Mountain railways 625
resorts hygiene 613
Mountains, attraction of, geodesy 526
in moon 523
physical geografy 551
Mounting specimens, collec. manuals 579
Mourning customs 393
Movables chattels 347
Movement cure 615
Movements of plants botany 581
Mowers agriculture 633
Mucilage manufacture 668
Muck agriculture 631
Mud architecture 693
Mudie's circulating libraries 027
Muggletonians sects 289
Mulberry agriculture 634
silk culture 638
Mule domestic animals 636
zoology 599
Multiple stars 523
Multiplex telegrafs commun. 654
Mummies 393
Mumps diseases 616
public health 614
Municipal bonds political economy 332
government, admin. 352
law 340
admin. 352
Munitions of war military eng. 623
Munster, Ireland history 941.9
travel 914.19
Mural decorations architecture 729
Murder criminal law 343
Muscle-reading 134
Muscular exercise hygiene 613.7
sense mental faculties 152
physiology 612
Museums and libraries 021
architecture 727
Museums, arrangement of specimens 579
art fine arts 708
medical 610
school 371
scientific natural science 507
Sunday opening 263
other museums see subject
Mushrooms botany 589
food 641
Music 780
box mech. instruments 789
copyright 655
ecclesiology 246
halls architecture 725
public worship 264
therapeutics 615
Musical form 781
instruments 780
acoustics 534
sounds " 534
Musicians' lives 927
Muskets see Rifles
Mustelidae zoology 599
Muster rolls 355
Mutilations of library books 024
war customs 399
Mutiny maritime law 347
Mutual aid societies pol. economy 334
insurance " 334
Mycenae ancient history 938
Mycology botany 589
Myopia eye diseases 617
Mysteries 133
Mystery plays Eng. literature 822
history 809
religion 244
theater 792
Mystic philosofers 189
Mysticism heresies 273
philosofy 149
Mystics medieval 189
Mythology, comparativ 291
German 293
Greek and Roman 292
Norse 293
Myths 291
Myxomycetes botany 589
Nahum Bible 224
Names of days of week 529
months 529
persons and places 929
Nantes, edict of religious history 272
Napkins customs 394
Naples, Italy history 945
travel 914.5

Napoleon, code law 349
Narcotic poisons therapeutics 615
Narcotics, action of hygiene 613
and stimulants ethics 178
Narrow-gage railways 625
Nasmith societies 267
Natal, Africa history 968
travel 916.8
Natatores birds 598
National aid to education 379
and state universities 378
" " state educ. 379
banks 332
churches *vs* organic unity 261
costumes 391
customs, special 390
debt finance 336
libraries 027
music 786
parks 711
politics 329
school supervision 379
soldiers' cemeteries 719
songs music 784
Nations, law of, international law 341
Nativ Americans political parties 329
Natural body doctrinal theology 233
gas chem. technology 662
economic geology 553
heating 697
history, Biblical 220.8
collectors' manuals 579
of countries 508.3
man 573
zoology 590
law 340
magic 133
philosofy 530
religion 210
science 500
bibliografy of 016.5
biografy of 925
selection evolution 575
stone building material 691
theology 210
Naturalists' lives 925
Naturalization 325
suffrage 324
Nature 500
law of 340
metaphysics 113
of being metaphysics 111
printing 655
Nautical astronomy 527
Naval architecture engineering 623
Naval battles 359
see also special countries
biografy 923
engineering 623
history 359
Amer. civil war 973.7
see also special countries
men, lives of 923
schools, e. g. Annapolis 359
science 359
ship-building 699
war 359
Navarre history 946
Navies 359
Navigation astronomy 527
laws 347
signals in, communicat. 654
useful arts 656
Navigators' lives 923
Navy administration 359
department, U. S. 353
Near sightedness medicin 617
schools 371
Nebraska, U. S. history 978.2
travel 917.82
Nebulae astronomy 523
Nebular hypothesis " 523
Necessity metaphysics 123
Necrology biografy 920
colleges 378
Necromancers 133
Needle-work 746
domestic economy 646
Negro education 371
race ethnology 572
nat. hist. of man 573
political standing 324
slavery 326
songs music 784
Nehemiah Bible 222
Neighborhood gilds social clubs 367
Nemean games Greek antiquities 913.38
Néo-Grec architecture 724
Neolithic age prehist. archeol. 571
Neology rationalism 211
Neoplatonism 186
Neoplatonists, apologetics against 239
Nepotism civil service 351
Neptune descriptiv astronomy 523
Nervous functions of animals 591
physiology 612
prostration 616
system, comp. anatomy 591
human " 611
Nestorian church 281

Netherlands history 949.2
travel 914.92
Nets for collectors biology 579
Neuralgia diseases 616
Neuroptera insects 595
Neuroses diseases 616
Neuters generation of animals 591
Neutrality international law 341
Nevada, U. S. history 979.3
travel 917.93
New Brunswick, Canada history 971
church, Swedenborgian 289
England geology 557.4
history 974
religious history 277.4
travel 917.4
France, Canada history 971
travel 917.1
Granada, S. A. geology 558.6
history 986
religious history 278.6
travel 918.6
Guinea geology 559.5
history 995
religious history 279.5
travel 919.5
Hampshire, U. S. history 974.2
travel 917.42
Hebrides, Australia history 993
Holland, Australia " 994
travel 919.4
Jersey, U. S. history 974.9
travel 917.49
Jerusalem church 289
Mexico, U. S. history 978.9
travel 917.89
Platonism 186
Plymouth colony history 974.4
South Wales " 994
Sweden, Penn. " 974.8
travel 917.48
Testament 225
biografy 225
Greek dialects 487
year customs 394
York, U. S. history 974.7
travel 917.47
Univ. of the State of 379
Newfoundland history 971
News rooms and reading rooms, free 027
Newspaper cuttings preservation 029
Newspapers ethics 179
general 070
history of journalism 070
Newt zoology 598
Ngami lake travel 916.7
Niagara Falls description 917.47
Nicaragua history 972.8
travel 917.28
Nicene, Ante-, church A. D. 100–325 281
councils religious history 270
creed 238
Post-, church A. D. 325–1054 281
Nickel coinage pol. econ. 332
inorganic chemistry 546
ores economic geol. 553
Nicobar islands history 992
Niello-work arch. decoration 729
Niger river description 916.6
Night-soil air pollution 614
Nihilism philosofy 149
socialism 335
Nile description 916.2
Nineveh ancient history 935
cuneiform lang. 492
Nirvana oriental philosofy 181
Nitrogen compounds air pollution 614
Nitroglycerin chemical tech. 662
Nobility heraldry 929.7
Nobles' lives 923
Nomads 397
Nomenclature, botany 580
theoretic chemistry 541
Nominalism philosofy 149
Non-Christian religions 290
and education 377
Non-Euclidean geometry 513
Non-jurors Eng. religious history 274.2
Nonconformists Eng. relig. hist. 274.2
sects 283
Normal schools 370
Norman architecture 723
conquests European hist. 940
Normandy, France history 944
Norse " 948
Normans, England " 942.02
Norrland, Sweden " 948.8
travel 914.88
Norse language 439
literature 839
mythology 293
Norsemen Scandinavian hist. 948
North Africa geology 556.1
history 961
religious history 276.1
travel 916.1
America antiquities 913.7
geografy 917
geology 557
history 970

North America languages 497
religious history 277
scientific travels and surveys 508.7
statistics 317
travel 917
American Indian literature 897
Atlantic states, U. S. geology 557.4
history 974
relig. hist. 277.4
travel 917.4
Carolina, U. S. history 975.6
travel 917.56
Central Africa geology 556.6
history 966
religious hist. 276.6
travel 916.6
states, U. S. geology 557.7
history 977
relig. hist. 277.7
travel 917.7
Eastern boundary, U. S. history 973.5
passage 919.8
Dakota, U. S. history 978.4
travel 917.84
pole travel 919.8
west boundary 973
passage travel 919.8
territory, U. S. hist. 977
travel 917.7
Northern lights aurora borealis 523
mythology 293
Northmen in America U. S. history 973.1
Scandinavian history 948
Norway botany 581.9
geology 554.8
history 948.1
religious history 274.8
travel 914.81
Norwegian language 439
literature 839
Romanesque arch. 723
Nosology medicin 616
Nostrums state control 614
Notaries public law 347
Notation, arithmetic 511
for book sizes 025
system of library classificat. 025
theoretic chemistry 541
Note-books literary methods 029
Notes, public and private, pol. econ. 332
Notion, concept. mental faculties 153
Nouvelle France history 971
Nova Scotia " 971
Novel reading ethics 175
Novel reading reading and aids 028
writing 808
Novelists' lives, English 928
Novels English literature 823
religious 244
Noxious materials, air pollution 614
vapors hygiene 613
Noxiousness of animals, economic zoology 591
Nubia, Egypt history 962
travel 916.2
Nuisances law 347
public health 614
Nullification in South Carolina 973.5
U. S. constitut. law 342
Number metaphysics 119
Numbering machines printing 655
Numbers, book of Bible 222
shelf library econ. 025
theory of algebra 512
Numismatics 737
Nuns and nunneries 271
Nursery domestic economy 649
food hygiene 613
lore folk-lore 398
Nurses physiological zoology 591
training of 610
Nursing domestic economy 649
medicin 610
Nusairian religion 297
Nutriments physiology 612
Nutrition physiological botany 581
zoology 591
physiology 612
Oases phys. geog. 551
Oaths ethics 179
legal 347
of allegiance pol. sci. 323
Obadiah Bible 224
Obedience doctrinal theology 234
Obelisks 718
Oberammergau passion play 792
Obituary addresses biografy 920
sermons 252
Object teaching education 372
Oblates monastic orders 271
Obligations social ethics 177
Obliquity of ecliptic to equator 525
Oboe musical instrument 788
Observation errors probabilities 519
mental faculties 152
Observations astronomy 524
auxiliary, astron. 522
Observatories 522
meteorologic 551

Topics in black-faced type are subdivided in tables.

Observatories, school apparatus 371
Observing powers education 372
Obstetrics 618
Occult sciences 133
Occultations and transits, astronomy 523
Occupations, injurious hygiene 613
mortality in public health 614
Ocean coast changes 551
currents phys. geog. 551
exploration 910
geografy maps 912
life zoology 591.9
mails 383
telegraf 654
tides 525
transportation sociology 387
useful arts 656
travel 910
Oceanica antiquities 913.9
geografy **919**
geology **559**
history **990**
religious history 279
scientific travels and surveys 508.9
statistics 319
travel **919**
Oceans physical geografy 551
Octateuch Bible 222
Odd Fellows secret societies 366
Odic force 134
Odontology animal 591
human anat. 611
Œcumenical councils, eccles. pol. 262
relig. hist. 270
Offensiv animals econ. zoology 591
materials, air pollu. 614
Offertory alms 264
organ music 786
sacred music 783
Office buildings, arch. 725
equipment and methods 651
tenure of civil service 351
training law education 340
Officers, army and navy lives 923
duties of public ethics 172
Official functions civ. serv. officers 351
register, U. S., civil service 351
residences architecture 725
Oficleide music 788
Ofidia reptils 598.1
Ofthalmia eye diseases 617
Ofthalmic surgery 617
Ofthalmoscope 617

Ohio, U. S., history **977.1**
travel 917.71
valley description 917.7
Oil, coal economic geology 553
manufacture 665
painting art 751
building 698
refining, air pollution 614
vegetable agriculture 633
Oils adulteration 614.3
chemical technology 668
Oklahoma, U. S. history 976.6
travel 917.66
Old Bactrian language 491
catholic church 284
Egyptian language 493
English " 427
French " 447
Norse " 439
literature 839
Prussian language 491
red sandstone geology 551
Saxon language 439
literature 839
Testament **221**
biografy 221
Oleomargarine, adulterations 614.3
chemical technology 664
Olfaction physiology 612
psychology 152
Oligarchy 321
Olive oil chemical technology 665
Olives agriculture 634
Olympiads chronology 529
Olympic games, Grecian antiquities 913.38
Omens delusions 133
Omnigraf **engraving** **760**
Omnipotence doctr. theology 231
Omnipresence, " " 231
Omniscience " " 231
Oneida community socialism 335
Ontogeny embryology 591
Ontology philosofy 111
Oolite geology 551
lithology 552
Oology ornithology 598
Oosporeae algae and fungi 589
Ooze lithology 552
Opal economic geology 553
Opening lib. and museums, Sunday 263
libraries on Sunday 024
Opera, amusement 792
dramatic music 782
ethics 175
houses architecture 725

Opera recitative vocal music 784
Operations, calculus of 517
Operativ surgery 617
Operetta music 782
Ophicleide " 788
Ophidia reptils 598.1
Ophthalmia eye diseases 617
Ophthalmic surgery 617
Ophthalmoscope 617
Opium habit ethics 178
hygiene 613
materia medica 615
traffic ethics 178
Opossums mammals 599
Oppositions solar system 523
Optics 535
Optimism philosofy 149
Optional studies curriculum 375
Oracles mind and body 133
Oral examinations education 371
Orange associations political soc. 363
Oranges agriculture 634
Orations English literature 825
Oratorios sacred music 783
Orators' lives, English 928
Oratory 808
American literature 815
English " 825
French " 845
German " 835
Greek " 885
Italian " 855
Latin " 875
sacred homiletics 251
Spanish literature 865
Orbits, descriptiv astronomy 523
earth " 525
theoretic " 521
Orchards agriculture 634
Orchestral music 785
sacred music 783
Orchestrion music 789
Orchids botany 584
fertilization of, botany 581
Ordeal, trial by law 340
Orders, holy ministry 262
sacraments 265
monastic 271
of architecture arch. design 729
knighthood heraldry 929.7
Ordinal church ritual 264
Ordinances " " 264
religious 265
Ordination eccles. polity 262
sacraments 265
Ordination sermons 252
Ordnance engineering 623
Ore deposits 553
dressing mining eng. 622
Oregon, U. S. history 979.5
travel 917.95
question U. S. hist. 973.5
Ores econom. geol. 553
metallurgy 669
mineralogy 549
mining 622
Orfan asylums 362
Organ music 786
sacred music 783
Organic and inorganic matter, differences 577
chemical industries 660
chemistry 547
remains paleontology 560
Organization central government, foreign states 354
city government 352
civil service 351
cooperation 334
courts law 347
laboring classes 331
schools education 371
Organografy botany 581
zoology 591
Oriental architecture 722
art 709
churches 281
customs 915
history 950
languages 490
literature 890
music theory 781
mythology 291
philosofy 181
question Turkey 949.6
religions 290
societies 490
Orientalia literature 890
Orientation ecclesiology 246
Origen early Christian philosofy 189
Origin of English language 422
evil natural theology 216
individ. soul metaphys. 129
knowledge " 121
language 401
life 576
man, natural hist. of man 573
religious belief 201
sexes evolution 575
species " 575

Topics in black-faced type are subdivided in tables.

Origin of speech philology 401
Original home of man ethnology 572
sin doct. theol. 233
Orinoco river travel 918.7
Orkneys, Scotland travel 914.1
Ornament, arch. dec. 729
Ornamental botany 716
design 745
gardening, landscape 710
Ornaments costume 391
prehistoric 571
toilet 646
Ornithology 598
Orology physical geografy 551
Orphan asylums 362
Orthodox church, holy or Greek 281
Orthodoxy heresies 273
Orthoepy comparativ philol. 411
English language 421
Orthografy English " 421
French " 441
German " 431
Greek " 481
Italian " 451
Latin " 471
Spanish " 461
Orthopedics surgery 617
Orthoptera articulates 595
Oryctology paleontology 560
Osmose of liquids 532
Ossetic language 491
Osteology, animal 591
human anatomy 611
Ostracism Greek antiquities 913.38
Ostracoda articulates 595
Ostrich zoology 598
farms agriculture 636
Otology anatomy of ear 611
Ottoman empire history 949.6
Out-buildings architecture 728
Outcast races 397
Outcrop geology 551
Outdoor relief charity 361
pauperism 339
sports 796
Outlawry criminal law 343
Outlines, compends, see subject
Outlining drawing 741
Outlooks landscape gardening 717
Outposts military science 355
Ovariotomy surgery 617
Over-hours hygiene 613
mantels fixt furniture 729
Overshot wheels mechanical eng. 621
Overstrain, physical hygiene 613.7
Overstudy school hygiene 371
Overture orchestral music 785
Overwork hygiene 613
mental mental hyg. 131
Owners, absentee and alien 333
liability of building 692
Ownership, land 333
Oxford, England travel 914.2
movement 283
university 378
Oxygen, chemistry 546
Oysters fisheries 639
food 641
preservd pub. health 614.3
zoology 594
Ozocerite economic geology 553
Ozone chemistry 546
hygiene 613
Pachyderms zoology 599
Pacific islands, Polynesia, hist. 996
travel 919.6
ocean physical geografy 551
slope, U. S. geology 557.9
history 979
states, U. S. history 979
religious hist 277.9
travel 917.9
Paddock lectures apologetics 239
Pædiatrics medicin 618
Pagan architecture 722
Paganism 290
Pageants customs 394
Paging machines printing 655
Pain sense physiology 612
Painless extinction of animal life 614
Paint, prehistoric 571
Painted decoration arch. dec. 729
Painters' lives 927
Painting fine art 750
house, sign, etc. 698
Paintings, collections of, galleries 708
engravings from 750
Paints chem. technology 667
Palaces architecture 728
of rulers arch. 725
Paleografy English language 421
inscriptions 417
Paleolithic age prehistoric arch. 571
Paleontology 560
Palestine ancient history 933
travel 915.6
Palestrina sacred music 783
Pali language 491
Palimpsests rare mss 091
Palingenesis metaphysics 129

Topics in black-faced type are subdivided in tables.

Palladium	inorganic chemistry	546
Pallium	sacred ornaments	247
Palmistry	magic	133
Palms	botany	584
Pamflet cases		025
Pamflets	**misc. collections**	040
	preservation of	025
on special topics see subject		
Panama	history	986
Pancreatic fluid	physiology	612
Pandects	Roman law	349
Panics, commercial	political econ.	332
Panjabi language		491
Panopticon	prison system	365
Panoramas	amusements	791
	painting	750
Pantheism	natural theology	212
	philosofy	147
Pantheon	architecture	722
Pantomimes	music	782
Papacy	eccles. polity	262
Papal infallibility	" "	262
Paper covers	library books	025
hanging	trade	698
manufacture		676
money		332
roofing	building	695
Papers on special topics, see subjects		
Papua	history	995
Papyrus	manuscripts	091
Parables	English literature	828
	New Testament	226
Paradise, garden of Eden,	Genesis	222
	primitiv home of race	572
	theology	236
Paraffin	chem. technology	665
Paraffining	building material	691
Paraguay	geology	558.9
	history	989
	religious history	278.9
	travel	918.9
Parallax, descriptiv	astronomy	523
practical	"	522
Parallelism	Scripture	220.8
Parasites, pathology	animals	591
	plants	581
Parasitic diseases	pathology	616
	public health	614
plants		581
Parcel post	post-office	383
Parchment	paper-making	676
Parent and child	law	347
Parents	ethics	173
Parian chronicle	Arundel marbles	481
chronicle	Greek history	938
Paris, congress of, 1856	intern. law	341
	expositions	606
Parish care of sick, fallen, etc.		258
	ecclesiastical polity	262
	law	352
	libraries	027
	missions theology	269
	work, societies for	256
Parks, public	landscape gardening	711
	local government	352
Parliament	constitutional law	342
	England	328
	houses of arch.	725
Parliamentary	blue books	328
	control admin.	351
	debates	328
	law	328
	reform England	328
Parlor magic		793
	pastimes	793
Parochial libraries		027
	schools and libraries	257
	education	377
Parodies	English literature	827
Parquetry	floors and flooring	721
Parricide	criminal law	343
Parrots	zoology	598
Parsee dualism	Christian heresies	273
Parseeism	Parsees	295
Parsing	English grammar	425
Parsis		295
Part songs	vocal music	784
Parthenon	architecture	722
Parties, political		329
	social	793
Partizan warfare	military science	355
Partnership	law	347
Parturition	obstetrics	618
Party conventions,	polit. science	329
feeling in religion, use of		280
Parzival	legends	398
Passion music		783
plays	English literature	822
	general history	809
	Oberammergau theat.	792
	religion	244
Passionists	monastic orders	271
Passions	philosofy	157
Pastel drawing		741
Pasteurism	protectiv inoculation	614
Pastoral letters		252
	theology	250
	visitations	253
Pastors	ecclesiastical polity	262
Patagonia	geology	558.2

Patagonia history 982
religious history 278.2
travel 918.2
Patent indexes literary methods 029
law law 347
patents 608
medicins adulterations 614.3
state control 614
Patents useful arts 608
Pathologic anatomy, study of dis. 616
Pathology botany 581
medicin 616
mental 131
of labor 618
zoology 591
Patience ethics 179
Patriarchal institutions pol. science 321
Patriarchs of Bible lives 220.9
Patricide law 343
Patriotism ethics 172
Patriots' lives 923
Patristic philosofers' lives 921
philosofy 189
theology 230
Patronage ethics 177
Pauper asylums 362
labor 331
Pauperism 339
and illiteracy, pub. schools 379
Pavements road engineering 625
sanitation 628
Pawnbroking loan institutions 332
Peabody fund education 373
Peace and war ethics 172
jubilee Boston 780
society international ethics 172
Peaches agriculture 634
Pearls precious stones 553
Pears agriculture 634
Peasantry political economy 331
Peat economic geology 553
fuel chem. tech. 662
domestic econ. 644
Pedagogics 371
Pedestrianism 796
ethics 175
Pediatrics medicin 618
Pedigrees genealogy 929
Pedobaptism 265
Peerage heraldry 929.7
Pehlevi language 491
Pelagianism heresies 273
Pelasgian architecture 722
Pelew islands history 996
Peloponnesus ancient history 938
Peloponnesus modern 949.5
Penal colonies, criminal law 343
law 343
Penance theology 265
Pencils writing materials 652
Pendulum geodesy 526
physics 531
Peninsular war Spanish history 946
Penitentiaries 365
architecture 725
Penmanship 652
Pennsylvania, U. S. history 974.8
travel 917.48
Penology criminal law 343
Pens writing 652
Pensions, government 351
teachers' 371
Pentateuch Bible 222
People's palace self-education 374
party, U. S. politics 329
Perception 152
Percussion caps explosivs 662
instruments music 789
Perfection ethics 171
Perfectionists sects 289
Perfumery chemical technology 668
toilet 646
Peridotite lithology 552
Periodical clubs self-education 374
Periodicals, general **050**
on special topics see subject
Peripatetic philosofers 185
Perjury law 343
Perkinism cures 615
Perpetual motion 531
Persecutions theology 272
Perseverance of the saints 234
Persia ancient history 935
geology 555.3
modern history 955
religious history 275.5
travel 915.5
Persian architecture 723
ancient 722
language 491
philosofers 181
Person of Christ doctrinal theol. 232
Personal appearance costumes 391
ethics 177
equation astron. 522
hygiene medicin 613
liberty Eng. law 347
names etymology 929
narrativs autobiografy **920**
property English law 347

Personal religion 248
rights 347
taxation, political economy 336
Personality dogmatics 233
metaphysics 126
Perspectiv descrip. geometry 515
drawing 742
effect arch. design 729
Perspiration therapeutics 615
Persuasion logic 168
Perturbations earth 525
of heavenly bodies 521
Peru antiquities 913.8
botany 581.9
geology 558.5
history 985
religious history 278.5
travel 918.5
Peruvian ancient architecture 722
Peshito language 492
Pessimism philosofy 149
Pestalozzian system of education 371
Pestilences public health 614
Pests agriculture 632
Petards military engineering 623
Peter, St epistles 227
Petition, right of Eng. const. hist. 342
Petitions administration 351
Petralogy lithology 552
Petrifaction, paleontology 560
Petrografy 552
Petroleum chemical technology 665
economic geology 553
Petrology geology 552
Peutingerian tables anc. geog. 913
Pew rents clerical support 254
Phallicism 291
Phanerogamia botany 582
Pharmacists' registration state control of 614
Pharmacopœial drugs adulterations 614.3
Pharmacopœias materia medica 615
Pharmacy econ botany 581
materia med. 615
Phases of moon 523
Phenicia ancient history 939
Phenician architecture 722
language 492
philosofers 181
Phenomena nutritiv, physiology 612
vital 577
Phi Beta Kappa 371
Philadelphia, Penn. history 974.8
travel 917.48
Philanthropists' lives 923
Philanthropy ethics 177
Philately postage stamps 383
Philemon epistles 227
Philharmonics, societies 780
Philippians epistles 227
Philip's, King, war, New Eng. hist. 974
Philologists' lives 924
Philology 400
bibliografy of 016.4
biografy of 924
classical 470-489
Philoprogenitivness phrenology 139
zoology 591
Philosofer's stone alchemy 540
Philosofers' lives 921
Philosofic apparatus 530
classification of knowl. 112
instrument making 681
magazines 105
systems 140
Philosofy 100
bibliografy of 016.1
biografy of 921
natural 530
of special topics see subject
Phoenicia ancient history 939
Phoenician architecture 722
language 492
philosofers 181
Phonetic shorthand 653
spelling English language 421
Phonics elementary education 372
Phonograf acoustics 534
Phonografy 653
Phonology 414
Phosfates as fertilizers agricul. 631
econ. geology 553
Phosforescence 535
Phosforescent animals 591
Phosforus compounds air pollut. 614
inorganic chemistry 546
materia medica 615
Photo-electrotyping 777
Photo-engraving 777
Photografic chemistry 771
Photografs, collections of 779
of animals in motion
art anatomy 743
zoology 591
heavenly bodies 523
Photografy 770
astronomy 522
optics 535
Photolithografy 775
Photometric observations stars 523

Photometrics optics 535
Photometry astronomy 522
Photozincografy 776
Phrase books English language 428
Phrenologists' lives 920
Phrenology 139
Phthisis lung diseases 616
 public health 614
Phyllopoda zoology 595
Phylloxera agriculture 632
Physical astronomy 523
 capacity of women educ. 376
 conditions mental charac. 136
 education hygiene 613.7
 school " 371
 geografy 551
 geology 551
 sciences 500
 units physics 530
Physicians, ethics 174
 lives of 926
 registration 614
Physics 530
 molecular 539
Physiognomy 138
Physiografy 551
Physiological acoustics 534
 botany 581
 chemistry man 612
 animals 591
 optics 535
 zoology 591
Physiology 612
 animal 591
 mental 131
 vegetable 581
Phytografy descriptiv botany 580
Piano and organ 786
Piccolo musical instrument 788
Pictorial art fine arts 700
Picts ancient history 936
 Scottish " 941
Picture galleries 708
Picturesque beauty esthetics 701
 views and tours 910
Pierce, F., presidency of 973.6
Piers arch. construction 721
 design 729
 bridges 624
 river and harbor eng. 627
Pietists heresies 273
Piety, practical 240
Pigeon holes office equipment 651
Pigeons zoology 598
Pigments chemical tech. 667
Pigments inspec. of, adulterations 614.3
Pigmies, human 573
Pigs domestic animals 636
Pile driving 624
Piles diseases 616
 foundations 721
Pilgrimages religious 248
Pilot laws 347
Pilots protection of life 614.8
 seamanship 656
Pineapple agriculture 634
 botany 584
Pines " 585
Pinnipeds mammals 599
Pioneers' lives 923
Pipes, elephant American antiquities 913.7
 hydraulics 532
 prehistoric archeology 571
 sewer 628
 steam heating 697
 waterworks 628
Piracy history 910
 law 343
Pirates' lives 910
Pisces zoology 597
Pisciculture business 639
Pistols manufacture 683
Pit dwellings prehist. archeology 571
Pitcairn Island history 997
 travel 919.7
Placers economic geology 553
Places, names of etymology 929
Plagiarism literary methods 029
Plague diseases 616
Plagues, history of public health 614
Plains, physical geografy 551
Plan architec. design 729
Plane geometry 513
 perspectiv 515
 trigonometry 514
Planetarium descript. astronomy 523
Planetary system, constitution of 523
Planets descript. astron. 523
 theoretic " 521
Planing machines mechanical eng. 621
Planisfere astronomy 522
Plans for building 692
 libraries library buildings 022
 of cities maps 912
Plantagenets English history 942.03
Plantation songs 784
Planting agriculture 631
Plants 580
 and animals differences 577
 classification of 580

Plants, gardening	716
geologic agents	551
landscape gardening	710
nature-printing from	655
paleontology	561
Plaster building material	691
casts sculpture	730
of Paris agriculture	631
art	731
Plastering masonry	693
Plate carving	736
domestic economy	642
glass insurance	368
manufacture	666
manufacture	671
Plated ware "	671
Plating books, library economy	025
Platinum inorganic chemistry	546
Plato ancient philosofy	184
Greek literature	888
idealism	141
Platonic philosofy	184
Platonists, new	186
Platt deutsch language	439
literature	839
Players' lives	927
Playing cards	795
Playright	655
Plays English drama	822
mystery history	809
religion	244
theater	792
parlor amusements	793
Pleading	347
Pleasure	157
Plectral stringd instruments	787
Pleuropneumonia cattle disease	619
diseases	616
Plotting surveying	526
Plow agriculture	631
Plumbing	696
house drainage	628
laws	696
Plurality of worlds astronomy	523
Plutology polit. econ.	330
Plutonic rocks lithology	552
Pneumatic tubes communication	654
Pneumatics	533
Pneumatology metaphysics	128
Pneumonia diseases	616
Poetical books of Bible	223
selections, Eng. literature	821
Poetry, Amer. civil war	973.7
literature	811
English literature	821
Poetry, French literature	841
German "	831
Greek "	881
humorous Eng. lit.	827
Italian literature	851
Latin "	871
macaronic Eng. lit.	827
philosofy of	808
sacred hymnology	245
satirical Eng. lit.	827
Spanish literature	861
Poets' lives	928
Pointed architecture	723
Poisonous secretion animals	591
Poisons chemical analysis	543
effects of diseases	616
medical jurisprudence	340
sale of state control	614
therapeutics	615
vegetable economic botany	581
Poker amusements	795
ethics	175
Polabian language	491
Poland history	943.8
language	491
literature	891
partition of history	947
Russian history	947
travel	914.38
Polar regions, antarctic geology	559.9
history	999
relig. hist.	279.9
travel	919.9
arctic geology	559.8
history	998
relig. hist.	279.8
travel	919.8
Polariscope optics	535
Polarization of light astronomy	522
chemistry	544
optics	535
Polemic sermons	252
theology	230
Police, local gov't	352
measures administration	351
medical state control	614
sanitary administration	351
Polish chemical technology	667
Polish language	491
literature	891
Polishing painting	698
Politeness ethics	177
etiquet	395
Political associations	363
conventions	329

Topics in black-faced type are subdivided in tables.

Political economists' lives 923
economy 330
ethics 172
history Amer. civil war 973.7
organizations 363
parties 329
science 320
sermons 252
status of woman 396
Politicians' lives 923
Politics 329
morals of 172
Polity, ecclesiastical 262
Polk, J. K., presidency of 973.6
Polka piano music 786
Poll-tax political economy 336
Pollen fertilization 581
Pollution, air, public health 614
water, by cemeteries 614
supply of towns 628
Polo amusements 797
Polonaise piano music 786
Polyandry customs 392
ethics 173
Polychromatic ornament 747
Polyfonic school music 783
Polygamy criminal law 343
customs 392
ethics 173
hygiene 613
Mormonism 298
Polyglot Bibles 220.5
Polygrafy special libraries 080
see subject for topical collections
Polymorfism crystallografy 548
Polynesia geology 559.6
history 996
languages 499
religious history 279.6
travel 919.6
Polypetalae botany 583
Polyps zoology 593
Polytheism 290
Polyzoa zoology 594
Pomology agriculture 634
Pompeii description 913.37
Pompey's pillar Egyptian antiquities 913.32
monuments 718
Ponds, artificial landscape garden. 714
Ponies students' honor 371
Pontificate, Roman eccles. polity 262
Poolselling ethics 175
Poor laws local government 352
pauperism 339
Poor rates 336
Poorhouses asylums 362
pauperism 339
Popery ecclesiastical polity 262
Popes church history 282
lives 922
Popham colony Maine history 974.1
Popish plot, 1679 English history 942.06
Popular ballads 784
delusions 133
education public 379
elections suffrage 324
errors 133
lectures self-education 374
science 500
summer schools self-educ. 374
Population statistics 312
Porcelain 738
chemical technology 666
painting 738
Pork packing public health 614
trichinae 614
Porosity properties of solids 539
Port Royal heresy 273
persecutions 272
Portents 133
Portland cement economic geology 553
manufacture 666
vase pottery 738
Porto Rico history 972.9
Portraits engraved 769
painting 757
Portugal geology 554.6
history 946.9
religious history 274.6
travel 914.69
Portuguese language 469
literature 869
Position, calculus of analyt. geom. 516
of body astron. 521
Positivism philosofy 146
Post-mortem examinations pathol. 616
Post-Nicene apostolic church 281
offices sociology 383
Postage stamps 383
Postal savings banks 332
service 383
telegraf 384
Potash chemical technology 661
Potato agriculture 633
disease agriculture 632
botany 581
Potters' clays, economic geology 553
Pottery art 738
manufacture 666
painting 738

Pottery	prehistoric	571
Poultry, diseases of		619
	domestic animals	636
Poverty	pauperism	339
Powder, baking	adulterations	614.3
	chemical technol.	664
gun	manufactures	662
manufacture		662
toilet		646
Power, transmission of		621
Practical astronomy		522
chemistry		542
mining		622
religion		240
Practice of medicin		616
Prairies	physical geografy	551
Praise meetings	public worship	264
Prakrit language		491
Prayer-books	church	264
church ritual		264
family		249
gage	natural theology	217
Lord's	gospels	226
meetings		264
natural theology		217
of Manasses, Apocryfa		229
private	personal relig.	248
Preachers' lives		922
Preaching		251
Preadamites		573
Precedence	heraldry	929.7
Precession	practical astronomy	522
	theoretic "	521
Precious metals	metallurgy	669
	mineralogy	549
stones	economic geology	553
	mineralogy	549
Predestination	doctrinal theology	234
Predicting machines	tides	525
Predictions		133
Preemption	state land	336
Preexistence of the soul		129
Prefixes	English etymology	422
Pregnancy	obstetrics	618
	pathology of	618
Prehistoric archeology		571
man, nat. hist. of man		573
Prelacy	ecclesiastical polity	262
Preordination, doctrinal theology		234
Preparatory schools, private		373
Preraphaelitism		759
Presages		133
Presbyter	ministry	262
Presbyterian orders in ministry		262
Presbyterians		285
Presbyterians	lives	922
Presbytery	ecclesiastical polity	262
Prescience	divine attribute	231
Prescription-writing	materia medica	615
Presents, Christmas	customs	394
Preservation of food		664
Preservativs	building	691
	zoolog. collections	579
Preserving	cookery	641
President	U. S. administration	353
	inauguration, ceremonies	394
	veto power	353
Presidential elections	U. S.	324
Presidents' lives		923
Press clippings		029
hydraulic	engineering	621
	science	532
journalism		070
liberty of		323
morals of		179
printing		655
public order		351
writing for	literary methods	029
Pressure gages	pneumatics	533
of air	meteorology	551
blood	physiology	612
clothing	hygiene	613
liquids	hydraulics	532
sense	physiology	612
Prestidigitation		791
Pretenders	English history	942.07
Prevention of contagious diseases		614
	fire, protection of life	614.8
Priapus	mythology	291
Prices of books	library admin.	025
political economy		338
Pride	ethics	179
Priest	ministry	262
Priesthood	Christian ministry	262
Priests, celibacy of		254
	ethics	176
lives, Roman cath.		922
Primacy	ecclesiastical polity	262
Primaries	suffrage	324
Primary schools	methods	372
	state system	379
Prime meridian	cosmic time	529
Primers	elementary education	372
Primeval man, natural hist. of man		573
	prehis. archeology	571
Primitiv apostolic church		281
Christianity		270
church		281
culture		571

Topics in black-faced type are subdivided in tables.

Primitiv man, customs of 571
Primogeniture, law 347
Primrose league, political associations 363
Prince Edward Is., Canada, history 971
Printed catalogs lib. economy 025
Printers' lives 926
marks 655
occupation hygiene 613
Printing 655
calico chemical tech. 667
inks " " 667
library administration 025
telegrafs, communication 654
trade 655
Prints, books of 769
cloth chemical technology 667
Prison associations 365
contracts polit. economy 331
disciplin 365
life Amer. civil war 973.7
Prisoners of war international law 341
Prisons 365
and reformatories arch. 725
hygiene 613
Private banks 332
grounds landscape garden. 712
international law 341
judgment ecclesiastical pol. 262
libraries 027
ownership of land 333
reading self-education 374
schools 373
girls 376
theatricals 793
ethics 175
vs public schools 379
worship 248
Privateers 341
Privately printed books rarities 094
Privileges, special, to library readers 024
to readers library admin. 025
Prize fighting, ethics 175
Prizes in school education 371
international law 341
maritime " 347
Pro-slavery slavery 326
Probabilities, mathematics 519
Probate law 347
Processes, building materials 691
Processions customs 394
Produce agriculture 633
Production political economy 338
Productions of the soil agriculture 633
Profanity ethics 177
Profecy Bible inspiration 220.1
Professional ethics 174
Professorships of books and reading 028
Profetic books of Bible 224
Profit sharing political economy 331
Profits capital and labor 331
Prognostications 133
Progress civilization 901
of population, statistics 312
sociology 301
Progressiv develop. of man evol. 575
Prohibited books book rarities 098
Prohibition party political science 329
temperance 178
Projectils military engineering 623
physics 531
Projections analytic geometry 516
descriptiv " 515
map geodesy 526
Promise ethics 174
Promissory notes law 347
political economy 332
Promotions of scholars 371
teachers 371
Pronunciation Eng. language 421
Proof and copy lit. methods 029
legal law 347
reading printing 655
Propagation of contagious diseases 614
Propellers ships 621
Proper names persons and places 929
Property capital 331
church 254
city local taxat. 352
law of title to 347
literary publishing 655
state finance 336
suffrage 324
tax 336
theory of, polit. econ. 335
Proportion, in art esthetics 701
plan arch. design 729
Proprietary libraries 027
Proscription criminal law 343
Prose composition Greek 488
Latin 478
Pro-slavery slavery 326
Prosody comparativ philol. 416
English language 426
French " 446
German " 436
Greek " 486
Italian " 456
Latin " 476
Spanish " 466
Prospecting mining engineering 622

Topics in black-faced type are subdivided in tables.

Prostitution administration 351
ethics 176
hygiene 613
Protection of animals 591
human life from accidents 614.8
political economy 337
Protectiv inoculation, pub. health 614
Protectorate English history 942.06
Protestant episcopal church 283
episcopalians' lives 922
sects, continental 284
Protestantism 284
and Romanism 282
Protofyta botany 589
Protoplasm animal physiology 591
biology 576
physiology 612
Protozoa paleontology 563
zoology 593
Provençal language 449
literature 849
Romanesque architecture 723
Proverbs, book of Bible 223
folk-lore 398
Providence, Divine theology 231
natural " 214
Provincialisms English language 427
Pruning agriculture 634
Prussia history 943.1
travel 914.31
Prussian language, old 491
Prussic acid toxicology 615
Psalmody music 783
Psalms Bible 223
Psalter church ritual 264
Pseudo gospels 229
Pseudomorfs crystallografy 548
Pseudoneuroptera articulates 595
Pseudonyms bibliografy 014
Psychic phenomena 134
research mind and body 130
Psychologic medicin 131
Psychology 150
Psychomancy divination 133
Pteridofyta botany 587
Pteropoda zoology 594
Public administration 350
analysts adulterations 614.3
annuities 336
baths 613
buildings, architecture 725
local government 352
charities 361
credit political economy 336

Public customs 394
debt finance 336
documents 328
print'g and dist'b'n 655
education 379
finance 336
gardens 711
health 614
local gov't 352
lands 336
law 340
libraries 027
sanitation 614
schools education 379
securities, polit. economy 336
shows ethics 175
customs 394
speaking, oratory 808
spirit patriotism 172
vs private schools 379
works administration 352
worship 264
Publishers' author catalogs 015
circulars 015
lives 920
Publishing 655
Pueblos dwellings 571
Puerperal diseases obstetrics 618
Pugilism amusements 796
ethics 175
Pulpit clergy 262
eloquence homiletics 250
oratory 251
Pulpits architec. accessories 729
sacred furniture 247
Pulse physiology 612
Pump pneumatics 533
Pumping engines mechan. eng. 621
waterworks 628
Pumps engineering 621
Punctuation English language 421
Punishment, future theology 237
law 343
school 371
Purgatory theology 236
Puritanism theology 285
Puritans English history 942.06
religious hist. 274.2
Mass. history 974.4
persecutions by 272
Purpose intention ethics 171
Pus medicin 617
Puseyism theology 283
Pushtu language 491
Puzzles 793

Topics in black-faced type are subdivided in tables.

Pyemia	diseases	616
Pygmies	folk-lore	398
	human	573
Pyramids	Egyptian	913.32
Pyrenees mountains	travel	914.6
Pyrometry	temperature	536
Pyrotechnics	chemical tech.	662
Pyrrhonism	Greek philosofy	186
Pythagorean philosofy		182
Pythias, knights of	secret societies	366
Quackery	delusions	133
	medical, state cont. of	614
Quadrant	astron. instrument	522
Quadrature	integral calculus	517
	of circle geom.	513
Quadrupeds		599
Quadruplex telegrafs	commun.	654
Quakers	lives	922
	persecutions	272
	sects	289
Qualifications of readers	lib. econ.	024
	teachers, education	371
	voters suffrage	324
Qualitativ analysis	chemistry	544
Quality	metaphysics	111
Quantics	algebra	512
Quantitativ analysis	chemistry	545
Quantity	English prosody	426
	metaphysics	119
Quarantine	administration	352
	contagious diseases	614
	laws, local admin.	352
Quarries, prehistoric		571
Quarrying	mining engineering	622
Quarrymen's occupation	hygiene	613
Quartermasters	military admin.	355
Quartet	sacred music	783
Quartets	vocal music	784
Quartz	lithology	552
	mineralogy	549
Quaternary age	geology	551
Quaternions	mathematics	516
Quays	harbor engineering	627
Queen Anne architecture		724
Queens' lives		923
Questions	see subject	
Quicksilver	see Mercury	
Quietism	heresies	273
Quietists, persecution of		272
"Quincy system,"	education	371
Quinin	therapeutics	615
Quoits	amusements	796
Quotations	English literature	828
	general collections	808
Rabbinic learning	Judaism	296
Rabbits	domestic animals	636
Race horse	amusements	798
	influence on mind	136
	mortality public health	614
	suffrage	324
Races of man		572
Racing	amusements	798
	ethics	175
Radiant energy of sun		523
	matter	539
	points meteors	523
Radiates	paleontology	563
	zoology	593
Radiation	heat	536
Ragged schools		377
Rags	public health	614
Railroad accidents		656
	corporations	385
	electric	621.3
	engineering	625
	horse	625
	law	385
	locomotivs	621
	officials' lives	923
	passenger stations, arch.	725
	reports, state	385
	transportation business	656
	sociology	385
	travel protect. of life	614.8
Railroading		656
Rails	railroad engineering	625
Railway guides		656
Rain	meteorology	551
Raising funds for libraries		021
Ram, hydraulic	engineering	621
	water hydraulics	532
Ranch, cattle		636
Range	domestic economy	643
Rank	heraldry	929.7
Rape	law	343
Rare binding	book rarities	095
	illustrations "	096
	printing "	094
Rarissima	"	090
Rat	noxious animal	591
	zoology	599
Rationalism	natural theology	211
Rationalistic age	religious history	270
Rationalists, apologetics against		239
Raw materials, duties on		337
Readers' aids		028
	elementary, Eng. language	428
	librarian's relations to	025
	qualifications lib. econ.	024
	selections Eng. lang.	428

Reading and aids 028
art of 808
books, higher, Eng. 828
clubs self-education 374
desks office equipment 651
elementary education 372
private self-education 374
proofs printing 655
rooms, free, library econ. 027
library buildings 022
Ready-reckoners 658
Reagents chem. anal. 543
Real estate law 347
taxation polit. econ. 336
presence, theology 265
Realism philosofy 149
Reapers agriculture 633
Reason in animals zoology 591
intuitiv faculty 156
natural theology 210
Reasoning faculty 153
Rebellion, political science 323
Southern U. S. hist. 973.7
Receipts medicin 615
useful arts 602
workshop, chem. tech. **660**
on special subj. see subj.
Recipe books cookery 641
Reciprocity free trade 337
Recognition hereafter doct. theol. 237
Reconstruction U. S. history 973.8
Recreation buildings arch. 725
hygiene 613.7
Recreations school hygiene 371
Recreativ arts **790**
Recruiting military science 355
Red Cross societies 361
river settlement, Canada, hist. 971
travel 917.1
Redemption theology 234
Redemptorists monastic orders 271
Reed wind instruments 788
Reference books library admin. 025
use of aids 028
department lib. econ. 025
exact literary methods 029
use of libraries 024
Reflecting circle, astron. instruments 522
instruments optics 535
telescopes astron. instr. 522
Reflection heat 536
light 535
mental faculties 153
Reflections perspectiv 515
Reflex action zoology 591
Reform bill, 1832 English suffrage 324
civil service 351
schools 364
education 371
sociology 301
spelling English language 421
Reformation, age of European hist. 940
religious history 270
Reformatories architecture 725
hygiene 613
temperance 178
Reformatory associations 364
Reformd church, Dutch 284
in Amer. 285
Reformers' lives 923
Refracting telescopes 522
Refraction heat 536
optics 535
practical astronomy 522
Refrigerators domestic economy 643
Refuge, houses of 364
Regeneration, theology 234
Regicides' lives 923
Regimen hygiene 613
Regiments, history of 353
U. S. civil war 973.7
Registers, family 929
of births, etc., public 312
deeds law 347
U. S. army officers 353
Registration public health 614
statistics 311
Regulation of med. prac., pub. health 614
Regulations, army 355
navy 359
readers library econ. 024
Regulators see special subject
Reincarnation 129
Relations, domestic, political science 323
foreign " " 327
of capital and labor 331
church and state, ethics 172
Relationship 929
Relativ location library economy 025
Relics miracles 231
Relief decoration, architecture 729
Religion **200**
and education 377
science apologetics 239
nat. theol. 215
bibliografy of 016.2
biografy of 922
Christian **200**
comparativ 290
natural **210**

Topics in black-faced type are subdivided in tables.

Religion, practical 240
Religions, non-Christian 290
Religious anecdotes 244
architecture 726
art 246
associations, Bible, etc. 206
Y. M. C. A., etc. 267
biografy 922
ceremonies sacraments 265
controversy 270
on special subjects see subject
education 377
of clergy 207
festivals 264
fiction 244
history 270
institutions 260
liberty 261
orders, monastic 271
painting art 755
ecclesiology 247
poetry 245
societies 206
of special sects 280
on special subjects see subject
symbolism 246
Remains, paleontology 560
Remedies, legal 347
pathology 616
therapeutics 615
Renaissance architecture 724
European history 940
fine arts 709
religious history 270
Rent of state property finance 336
political economy 333
Rents, pew clerical support 254
Repentance doctrin 234
Reporting journalism 070
shorthand 653
Reports, law, American 345
British 346
library administration 025
see subject
Repoussé work fine arts 739
Representation 324
Representativ gov't polit. sci. 321
Reprisal international law 341
Reproduction, botany 581
physiology 612
zoology 591
Reproductiv power, mental faculties 154
Reptils paleontology 568
zoology 598.1
Republic political science 321
Republican institutions 321
party, U. S. 329
Repudiation finance 336
Reredos sacred furniture 247
Rescue protection of life 614.8
Reservoirs 628
Residences architecture 728
official, arch. 725
Residual calculus 517
Resins chemical technology 668
fossil, economic geol. 553
Resistance of materials, engineering 620
molec. phys. 539
water hydraulics 532
Resonators acoustics 534
Resorts, health hygiene 613
Respiration animals 591
physiology 612
Respiratory system, anatomy 611
Responsibility, human doct. theol. 233
ethics 171
Rest and sleep, hygiene 613.7
Restaurants 642
architecture 725
Restoration of the Jews 296
Restriction of contagious diseases 614
Resumption U. S. finance 336
Resurrection Christology 232
doctrinal theology 236
Resuscitation of the drownd, protect. of life 614.8
Retreats, religious 269
Retribution doctrinal theology 237
Revelation, Book of Bible 228
divine 231
Revenue state finance 336
Reviews, see subject reviewd
Revised statutes American law 345
version Eng. Bible 220.5
New Test. 225
Old " 221
Revival, Gothic architecture 724
Grecian " 724
of letters, literary history 809
Romanesque arch. 724
Revivals music 783
of religion 269
Revolution, American history 973.3
English, 1688 " 942.06
of the earth astronomy 525
Revolutions political science 323
Rewards and punishments theology 233
school 371
Rhaeto-Romansch language 459
Rhetoric 808

Topics in black-faced type are subdivided in tables.

Rhetoric, sacred homiletics 251
Rheumatism diseases 616
Rhine travel 914.3
Rhinoceros zoology 599
Rhizopoda zoology 593
Rhode Island, U. S. history 974.5
travel 917.45
Rhodes ancient history 939
knights of 929.7
modern history 949.9
Rhymes English prosody 426
Rhyming dictionaries Eng. lang. 426
Rhynchota insects 595
Richard 1 English history 942.03
2 " " 942.03
3 " " 942.04
Riddles amusements 793
popular life 398
Riding 798
Rifle manufacture 683
military science 355
shooting amusements 799
Rifled ordnance, military engineering 623
Right of search 341
standard of ethics 171
Righteousness, personal religion 248
Rights and liberties, const. law 342
land political economy 333
political science 323
Rinderpest veterinary medicin 619
Rings, finger costume 391
sacred ornaments 247
Rio Janeiro history 981
travel 918.1
Riots law 343
in any locality see name of place
Ripple-marks geology 551
Rites and ceremonies customs 394
Ritual public worship 264
Ritualism 264
River engineering 627
transportation sociology 387
Rivers physical geografy 551
water supply 628
Rivets machine tools 621
Roadbed railroad engineering 625
Roads and pavements engineering 625
elevated " 625
sociology 388
engineering 625
military military eng. 623
Robbery criminal law 343
Robin Hood English ballads 821
legends 398
Rock analysis chemistry 543
Rock salt economic geology 553
Rocks lithology 552
Rocky mountains, U. S. travel 917.8
Rodentia zoology 599
Roland legends 398
Rolls of honor military 355
schools 371
Romaic language 489
literature 889
Roman antiquities 913.37
architecture 722
art 709
catholic church 282
catholics' lives, collected biografy 922
history, ancient 937
inquisition, persecutions 272
law 349
literature 870
mythology 292
orders monastic orders 271
sculpture 733
Romance, English 823
languages 479
religious 244
Romanesque architecture 723
revival, architecture 724
Romanism 282
Romans epistle Bible 227
Rome, ancient history 937
church of 282
missions 266
theology, doctrins 230
modern 945
travel 914.5
Romish church 282
Rommany gipsies 397
Rood screen architecture 729
sacred furniture 247
Roof construction architecture 721
coverings building 695
Roofs architectural design 729
engineering 624
Rope making 677
Rose botany 583
culture 716
Roses, wars of, English history 942.04
Rosetta stone, Egyptian inscriptions 493
Greek " 481
Rosicrucians secret societies 366
Rotary blowers, engineering 621
pumps " 621
Rotation of earth 525
heavenly bodies 521
sun 523

Rotation physics 531
Rote teaching education 371
Rotifera zoology 595
Roulettes curves, anal. geom. 516
Roumania, Turkey history 949.8
travel 914.98
Roumanian language 459
Round dances, amusements 793
ethics 175
table Arthurian romances 398
Roundheads English history 942.06
Rowing sport 797
Royal houses heraldry 929.7
lives collected biografy 923
Rubber manufactures 678
Rubies economic geology 553
Rubrics Anglican ritual 264
Ruins antiquities 913
Rulers' lives collected biografy 923
palaces architecture 725
Rules for readers library econ. 024
staff " " 023
of order, parliamentary 328
school 371
Rum chemical technology 663
Rumania history 949.8
travel 914.98
Rumansh language 459
literature 859.9
Ruminants zoology 599
Runes 439
Anglo-Saxon 429
German 437
Gothic 439
Runic inscriptions 439
Running hygiene 613.7
sports 796
Rural architecture 728
cemeteries 719
economy agriculture 630
life " 630
sports 796
water supply 628
Russia botany 581.9
geology 554.7
history 947
in Asia history 957
religious history 274.7
statistics 314
travel 914.7
Russian America, Alaska history 979.8
architecture 723
baths hygiene 613
church sects 281
cyclopedias 037
Russian essays, collected 047
journalism 077
language 491
newspapers 077
periodicals, general 057
societies, general 067
Rust, anti-, processes, build. materials 691
Ruth Bible 222
Ruthenian language 491
Ryswick, peace of treaties 341
Sabbatarians Christian sect 289
Sabbath 263
schools 268
Sabeism philosofy 181
Sabellian heresies 273
Sacramentals public worship 264
Sacraments 265
ritual 264
Sacred art 246
biografy 922
drama miracle plays 792
religion 244
furniture 247
history 270
music 783
poetry 245
rhetoric 251
Sacrifice birth customs 392
Christology 232
Sacrifices Bible 220.9
sacraments 265
Sacrilege 265
Saddlery trade 685
Safes manufactures 672
Safety lamps mining engineering 622
Sagas, Norse 839
Sahara desert physical geografy 551
travel 916.6
Sail-making trade 689
Sailing directions 656
navigation 527
Sailors' lives 926
Saint Albans raid U. S. history 973.7
Vermont hist. 974.3
Bartholomew's day, persecution 272
Bernard, hospice of, prot. of life 614.8
Helena history 997
John of Jerusalem, knights of 929.7
Louis, Mo. history 977.8
travel 917.78
Petersburg, Russia history 947
travel 914.7
Simonism socialism 335
Saints doctrin 235
lives of **922**

Topics in black-faced type are subdivided in tables.

Salamanders batrachia 597
Sale catalogs author 018
dictionary 019
of books 017–019
subject 017
Salic laws German law 349
Salin waters economic geology 553
Salmon fishing, amusement 799
business 639
Saloons temperance 178
Salt, lithology 552
manufacture 664
rock economic geology 553
Salts, chemical technology 661
theoretic chemistry 541
Salvage maritime law 347
Salvation doctrinal theology 234
army 267
Samaritan language 492
Samuel Old Testament 222
San Francisco history 979.4
travel 917.94
San Salvador history 972.8
Sanctification doctrinal theology 234
Sanctuary legal antiquities 340
Sanctus sacred music 783
Sand storms 551
Sandals costumes 391
Sands economic geology 553
Sandstone " " 553
lithology 552
old red 551
Sandwich Islands history 996
Sangreal legends 398
Sanguification, physiology 612
Sanhedrim Judaism 296
Sanitariums health resorts 613
Sanitary affairs public health 614
appliances hygiene 613
public health 614
commissions local gov't 352
U. S. 361
engineering 628
laws public health 614
measures " " 614
police administration 351
reports public health 614
science engineering 628
Sanitation, engineering 628
public 614
veterinary, pub. health 614
Sanskrit language 491
literature 891
Santo Domingo history 972.9
Saracen empire history 953
Saracenic architecture 723
Sarcofagi burial customs 393
Sardinia island, modern 945
Sassanian architecture 722
Satan doctrinal theology 235
Satellites theoretic astronomy 521
Satire American literature 817
English " 827
French " 847
German " 837
Greek " 887
Italian " 857
Latin " 877
Spanish " 867
Satirical music, dramatic 782
Satirists' lives, English 928
Saturn astronomy 523
Saurians 598.1
Sauropterygia, reptils 598.1
Savings banks, political econ. 332
Savoy, declaration doct. theology 238
Saw mills 674
Saws tools 621
Saxon, Anglo-, language 429
literature 829
old language 439
literature 839
Romanesque architecture 723
Saybrook platform creeds 238
Scalds surgery 617
Scales, measuring instrument 530
music 781
practical chemistry 542
watch and instrument making 681
weights and measures 389
Scalping war customs 399
Scandal ethics 177
Scandinavia history 948
Scandinavian cyclopedias 038
essays, collected 048
journalism 078
language 439
literature 839
newspapers 078
periodicals, gen. 058
philosofy 198
protestants 284
societies, gen. 068
Scarlet fever diseases 616
public health 614
Scavenging of towns 628
Scene painting painting 750
theater 792
Scepticism ancient philosofy 186
heresies 273

Topics in black-faced type are subdivided in tables.

Scepticism natural theology 211
philosofic systems 149
Sceptics' lives, biografy 922
Schism eccles. history 270
polity 262
Schists lithology 552
Schizomycetes botany 589
Scholar's life, self-culture 374
Scholastic philosofers' lives 921
philosofy 189
School architecture 727
funds state education 379
furniture 371
hygiene education 371
medicin 613
libraries education 371
library economy 027
texts English language 428
Schools and libraries, library econ. 021
architecture 727
education 370
library library econ. 020
music mus. education 780
of painting 759
parochial education 377
parish work 257
public 379
reform 364
architecture 725
Sunday 268
see also special subjects
Science and religion apologetics 239
natural theol. 215
applied **600**
biografy of 925
elementary education 372
mental **150**
military 355
moral **170**
natural **500**
naval 359
of religion 201
political **320**
sanitary 628
social sociology **300**
study and teaching 507
Sciences, classification of 501
natural **500**
occult 133
Scientific buildings architecture 727
men lives of 925
surveys, general **508.3**
travel **508.3**
Scientists, apologetics against 239
Scope and usefulness of libraries 021
Scorpions zoology 595
Scotch churches, presbyterian 285
covenant creeds 238
English dialects 427
language 491
philosofers' lives 921
philosofy 192
presbyterians' lives 922
Scotland antiquities 913.4
botany 581.9
geology 554.1
history 941
religious history 274.1
travel 914.1
zoology 591.9
Scottish episcopal church 283
poetry 821
Scrap book methods 029
books general collections **040**
for special, see subject
Screening mining engineering 622
Screens arch. construction 721
fixt furniture 729
Screw cutting 621
physics 531
propellers engineering 621
Script writing 652
Scriptural biografy Bible 220.9
Scripture chronologies 220.9
commentaries 220.7
concordances 220.2
dictionaries 220.3
harmonies of gospels 226
histories 220.9
readings public worship 264
Scriptures, holy Bible **220**
Scrofula diseases 616
Scroll-sawing 749
Sculptors' lives 927
Sculpture **730**
prehistoric 571
sacred ornamentation 247
Scurvy diseases 616
Scutage feudalism 321
Scythian language 494
literature 894
Sea anemones zoology 593
cucumbers " 593
fishery trade 639
laws 347
mosses algae 589
serpent delusions 133
shore resorts, hygiene 613
urchins zoology 593
weeds botany 589

Seal engraving		736
fishing		639
Seals	heraldry	929.8
	zoology	599
Seamanship		656
Seamen's lives		926
Search, right of	international law	341
Seasickness		616
Seaside cottages	architecture	728
resorts	hygiene	613
Seasons	astronomy	525
	hygiene	613
Seating capacity	architecture	729
Seats, country	"	728
Secession	church of Scotland	285
	law	342
	U. S. history	973.7
Second advent of Christ		232
adventists	sects	289
sight		133
Secondary schools, private		373
public		379
Secret ballot	suffrage	324
service	Amer. civil war	973.7
societies		366
	college	371
writing	cryptografy	651
Secretion	animals	591
	physiology	612
Sects, Christian		**280**
Secular education		377
Sedentary life, hygiene		613
Sedimentary rocks	lithology	552
Sedition	law	343
See	eccles. polity	262
Seismology		551
Selection	evolution	575
Self culture		374
defense, art of	athletics	796
education		374
government	polit. science	321
	school discip.	371
Sematology	philology	419
Seminaries	architecture	727
	education	373
	theological religion	207
for special topics see subject		
Seminary method of instruction		371
Semitic languages		492
literatures		892
Sensation, beginnings of	biology	576
in animals	zoology	591
plants	botany	581
mental faculties		152
Sensationalism		145
Sense	mental faculties	152
Sensibility	" "	157
Sensitiv plant	botany	583
	physiol. botany	581
Sepoy war, Indian	history	954
Septuagint	Bible	221
Sepulcher	cemeteries	718
Sepulture	customs	393
	public health	614
Serfs, Russian	history	947
slavery		326
Sergeants'-at-law	lives	923
Serials, see subject		
Sericulture		638
Sermons		252
	public worship	264
Serpent-charming		591
worship		291
Serpentine	economic geology	553
	lithology	552
Serpents	zoology	598.1
Servants	domestic economy	647
	ethics	173
	law	347
Servia	history	949.7
	travel	914.97
Servian language		491
Service books	church ritual	264
library		023
Servitude	slavery	326
Session laws,	American	345
	English	346
Seven champions		398
sages		398
sleepers		398
wonders of the world		609
Seventh day adventists	sects	289
baptists	"	286
Sabbath		263
Sewerage	city administration	352
	disposal of	628
	engineering	628
Sewing	domestic economy	646
	machine	687
	shoes	685
	tailoring	687
Sex customs		392
in education		376
nature		577
influence of	therapeutics	615
mortality	public health	614
origin of	evolution	575
production of	embryology	591
Sexes	mental characteristics	136
Sextant	astron. instrument	**522**

Topics in black-faced type are subdivided in tables.

Sexual ethics 176
science 612
selection evolution 575
variation animals 591
Sferical astronomy 522
geometry 513
projections, descr. geom. 515
trigonometry 514
Sfinx Egyptian mythology 299
Sfygmograf diagnosis 616
physiology 612
Shades and shadows, descrip. geom. 515
Shaft sinking, mining 622
Shafting mill work 621
Shakers sects 289
Shale lithology 552
Sharks 597
Shay's rebellion Mass. history 974.4
Sheep, diseases of, compar. medicin 619
domestic animals 636
Shelf dep't library economy 025
Shell banks prehist. archeology 571
fish business 639
zoology 594
Shells fossil 564
prehist. archeology 571
projectils 623
zoology 594
Shelter of laboring classes, pol. econ. 331
Shelving library building 022
Shemitic languages 492
Sheol eschatology 236
Sherifs law 347
local government 352
Shetland islands history 941
travel 914.1
Ship building 699
canals engineering 626
sociology 386
life hygiene 613
propulsion mechanical eng. 621.1
railways engineering 625
Shipping and freighting, useful arts 656
history of sociology 387
laws 347
Ships, iron 699
clad military eng. 623
transportation 656
Shipwrecks adventures 910
Shipwreckt, rescue of, prot. of life 614.8
Shoals, Isles of history 974.2
travel 917.42
Shock result of injury 617
Shoemaking manufacturing 685
Shoes costume 391
Shooting, sport 799
Shop girls labor of women 331
Shorthand 653
Shortsightedness eye diseases 617
Shows entertainments 791
public customs 394
Shrimps zoology 595
Shrubbery landscape gardening 715
Shrubs botany 582
Siam history 959
travel 915.9
Lower history 959
Siamese architecture 722
language 494
twins deformities 617
natural hist. of man 573
Siberia geology 555.7
history 957
religious history 275.7
travel 915.7
Sibyllin books magic 133
Sicilian Norman architecture 723
Sicily ancient history 937
modern " 945
Sick, food for hygiene 613
hospitals for 362
architecture 725
ministry of sacraments 265
parish care of 258
room domestic economy 649
Sidereal month astronomy 523
time chronology 529
Sieges military engineering 623
Sight anatomy 611
asylums for the blind 362
diseases of 617
mental faculties 152
optics 535
physiology 612
Sign boards advertising 659
language 419
painting 698
Signal service meteorology 551
Signals in navigation acoustics 534
military military engin. 623
telegrafic 654
Signatures printing 655
Sikh religion 299
Silesia, Austria history 943.7
Prussia " 943.1
Silhouets 741
Silk culture 638
dyeing chemical technology 667
manufacture 677
worm agriculture 638

Silos	agriculture	633
Silurian age	geology	551
Silver	inorganic chemistry	546
	metallurgy	669
	mining	622
	money	332
	ores economic geol.	553
	plating manufacture	671
Silversmiths' work	fine arts	739
Silverware	domestic economy	642
	fine arts	739
	manufactures	671
Simony	canon law	348
	eccles. polity	262
Sin	doctrin	233
Sinai	ancient history	939
	modern "	953
Sinaitic inscriptions		492
Singers' lives		927
Singing	public worship	264
	vocal music	784
Single standard	coinage	332
Sinking fund	finance	336
Sintooism		299
Sirocco	meteorology	551
Sisterhoods	parochial work	255
	religious orders	271
Size notation for books		025
Sizes of books, shelf dep't lib. econ.		025
standard	literary methods	029
Skating		796
Skeleton, human		611
leaves	fancy work	746
of animals	anat.	591
Skeletons, preparations of		579
Skepticism	natural theology	211
Sketching	drawing	741
Skild labor, political economy		331
Skill, games of, amusements		794
	ethics	175
Skirmishing	military science	355
Skull	anatomy	611
	craniology	573
Slander	ethics	177
Slang	English language	427
	French "	447
	German "	437
	Greek "	487
	Italian "	457
	Latin "	477
	Spanish "	467
Slater fund	education	373
Slates	building stones	553
	lithology	552
Slating	building	695
Slaughter houses	air pollution	614
	industrial sanit.	628
Slavery		326
abolition of, U. S. history		973.7
Slaves' biografy		326
Slavic cyclopedias		037
essays, collected		047
languages		491
periodicals, gen.		057
philosofy		197
Slavonic church language		491
Sleep		135
and rest	hygiene	613.7
of plants	botany	581
walking		135
Sliding friction	mechanics	531
Slight of hand		791
Sligo, Ireland	history	941.7
Slöjd	education	371
Sloth	ethics	179
Slovakian language		491
Slovenian "		491
Sluices	engineering	621
Smallpox	diseases	616
	inoculation, pub. health	614
	public health	614
Smell, nerves of	anatomy	611
	physiology	612
senses		152
Smelting	metallurgy	669
Smith college, colleges for women		376
Smithsonian institution		506
Smoke	air pollution	614
	flues heating	697
	nuisance, sanitation	628
Smoking	ethics	178
	hygiene	613
Smuggling	criminal law	343
Smut	agriculture	632
	fungi	589
Snakes		598.1
Snow	meteorology	551
Soap-making	air pollution	614
	chemical technology	668
Sociability	animals	591
Social clubs		367
customs and manners		394
democracy	socialism	335
distinctions	ethics	177
	rank	929.7
drink	temperance	178
ethics		177
evil		176
science		300
worship		264

Topics in black-faced type are subdivided in tables.

Socialism 335
Societies 360
debating 374
for home study 374
parish work 256
general 060
special see subject
Society, ethics 177
for ethical culture 170
houses student life 371
islands history 996
of Christian endeavor 267
Jesus monastic orders 271
the Cincinnati 369
Socinians heresies 273
unitarian church 288
Sociology 300
bibliografy of 016.3
biografy of 923
Socratic philosofy 183
Soda chemical technology 661
Sodalities parish work 256
Soil public health 614
Soils agriculture 631
chemical analysis 543
Solar engines engineering 621
spectrum astronomy 523
system descriptiv astron. 523
time chronology 529
Soldiers' cemeteries, national 719
homes institutions 362
lives biografy 923
military science 355
orfan schools 362
Sol-fa method music 784
Solids geometry 513
properties of physics 539
Solitary confinement punishments 343
Solitude 177
Solomon islands history 993
song of Bible 223
Solubility theoretic chemistry 541
Solution, chemical 542
Solutions quantitativ analysis 545
Somnambulism 135
Somnambulists' lives 920
Song festivals music 784
of Solomon Bible 223
the three children, Bible 229
Songbooks folk-lore 398
Songs and ballads, English literature 821
music 784
Sonnets English literature 821
Sophist philosofy 183
Sorcery 133
Sorghum agriculture 633
manufactures 664
Soteriology doctrinal theology 234
hygiene 613
Soudan, Africa, history 966
Egypt, " 962
Soul in animals 591
metaphysics 128
origin of metaphysics 129
Sound physics 534
South Africa geology 556.8
history 968
religious history 276.8
travel 916.8
America antiquities 913.8
description 918
geology 558
history 980
religious history 278
scientific travel and surveys 508.8
statistics 318
American Indian literature 898
languages 498
travel 918
Atlantic states, U. S. geology 557.5
history 975
religious history 277.5
travel 917.5
Carolina, U. S. history 975.7
travel 917.57
Central Africa geology 556.7
history 967
religious history 276.7
travel 916.7
states, U. S. geology 557.6
history 976
relig. hist. 277.6
travel 917.6
Dakota, U. S. history 978.3
travel 917.83
Sea islands history 996
travel 919.6
scheme stocks 332
west, U. S. history 978
travel 917.8
Southern states U. S. history 975
Sovereigns' lives 923
Sovereignty pol. science 320
Space astronomy 523
metaphysics 114
sense physiology 612
Spain antiquities 913.4
Arabs in history 946
botany 581.9

Topics in black-faced type are subdivided in tables.

Spain geology 554.6
history 946
religious history 274.6
travel 914.6
Spanish America history 972
travel 917.2
art 709
cyclopedias 036
essays, collected 046
journalism 076
inquisition, persecutions 272
language 460
literature 860
newspapers 076
periodicals, general 056
philosofy 196
societies, general 066
Spar economic geology 553
Spark, induction electricity 537
Sparring amusements 796
Sparta ancient history 938
Spas hygiene 613
Speakers oratory 808
Speaking tubes communication 654
Specie payment banking 332
Species, origin of 575
Specific customs taxation 336
gravity chemistry 542
of liquids physics 532
physics 531
Specifications for building 692
Specifics therapeutics 615
Specimens, arrange. in museums 579
Spectacles disorders of vision 617
optics 535
Specters 133
Spectroscope, optics 535
Spectroscopy, astronomy 522
Spectrum analysis astronomy 523
chemistry 544
optics 535
Speculation banks 332
ethics 174
Speech, impediments of 612
organs of acoustics 534
human anatomy 611
philology 400
phonology 414
physiology 612
visible comparativ philol. 414
Speeches English oratory 825
Spelling books, English language 428
phonetic, " " 421
reform " " 421
Sph- see Sf-

Sphagna botany 588
Spices adulterations 614.3
chemical technology 664
Spiders zoology 595
Spindles stone age 571
Spine, curvature of diseases 616
Spinning manufactures 677
Spires and domes, arch. construction 721
architectural design 729
Spirit doctrinal theology 233
leveling, geodesy 526
Spiritism 133
Spiritual body doctrinal theology 233
heredity " " 233
mental characteris. 136
metaphysics 129
Spiritualism metaphysics 111
modern 133
sects 289
Spirituous liquors chem. tech. 663
temperance 178
Spitzenberg history 998
travel 919.8
Spoils system, civil service 351
Sponges zoology 593
Spontaneity metaphysics 127
Spontaneous combustion chemistry 541
generation 576
Sports, outdoor 796
parlor 793
Springs, hot physical geology 551
mineral economic geol. 553
waterworks 628
Sprinkling of streets sanit. of town 628
Squares, least geodesy 526
probabilities 519
magic, arithmetic 511
tables of 510
Squaring the circle geometry 513
Squids zoology 594
Squinting eye diseases 617
Stables agriculture 636
air pollution 614
architecture 728
Stacks library economy 022
Staff library economy 023
military science 355
music 781
Stage ethics 175
theater 792
Staind glass 748
Staining chemical tech. 667
Stairbuilding building 694
Stalactites and stalagmites 551
Stammering 616

Topics in black-faced type are subdivided in tables.

Stamp act U. S. history 973.3
  laws finance 336
Stamps, postage post-office 383
Standard sizes literary methods 029
  time 529
Standing army 355
Star catalogs 523
  chamber Eng. history 942.05
  clusters 523
  drift 523
  fish 593
Starch adulterations 614.3
  manufacture 664
Stars descriptiv astron. 523
Starvation dietetic diseases 616
  hygiene 613
State and church church 261
  political science 322
  state ethics 172
  banks 332
  capitols architecture 725
  control of cities adminis. 352
  debts finance 336
  department, U. S. government 353
  education 379
  ethics 172
  governments 353
  laboratories adulterations 614.3
  law reports U. S. 345
  libraries 027
  medicin 614
  ministries of architecture 725
  monopolies 336
  ownership of land 333
  political science 320
  properties finance 336
  rights U. S. const. law 342
  socialism 335
  supervision of schools educ. 379
  trials criminal law 343
  universities colleges 378
  state education 379
Statecraft 320
States-general, French legislature 328
Statesmanship political science 320
Statesmen's lives 923
Static electricity 537
Statics 531
Stationary engines 621.1
Stationery paper manufacture 676
Statistical methods 311
Statistics 310
  labor 331
  library administration 025
  see special subjects
Statuary 730
Statute law American 345
  British 346
Steam boilers 621.1
  inspection, prot. of life 614.8
  engine mech. eng. 621.1
  physics 536
  fitting 696
  heating building 697
  engineering 621.1
  navigation sociology 387
  useful arts 656
  physics 536
  pipes 696
  pumps 621
  ships building 699
  transportation useful arts 656
Stearin organic chemistry 547
Steel building material 691
  engraving 762
  manufactures 672
  metallurgy 669
  structures arch. 721
  works " 725
Steeples architecture 729
Steering seamanship 656
Stellar astronomy 523
Stencils practical printing 655
Stenografy 653
Stephen English history 942.02
Stereometer specific gravity 531
Stereometry gaging 511
Stereopticon optics 535
Stereoscopes " 535
Stereotomy descr. geometry 515
Stereotyping printing 655
Stethometry physiology 612
Stethoscope diagnosis 616
Sties, hog air pollution 614
Stigmata devotional 248
Still-life painting 754
Stills chemical technology 668
Stimulants action of, hygiene 613
  and narcotics med. 615
  ethics of using 178
  manufacture 663
Stipple engraving 765
Stock exchanges banks 332
  jobbing 332
  live agriculture 636
Stockholm history 948.7
  travel 914.87
Stockings costume 391
  clothing 646
Stocks political economy 332

Topics in black-faced type are subdivided in tables.

Stoic philosofy 188
Stone age 571
and iron structures architec. 721
artificial, chemical technology 666
building materials 691
carving sculpture 736
construction masonry 693
cutting masonry 693
sculpture 731
stereotomy 515
remains 571
ware manufactures 666
Stonehenge English antiquities 913.4
Stones econ. geology 553
Storage of electricity, engineering 621
science 537
rooms library economy 022
Stores, cooperativ political econ. 334
Stories English fiction 823
Storms, electric 538
meteorology 551
Stoves domestic economy 644
heating building 697
manufactures 672
Strains bridge engineering 624
Strategy 355
Stratification, geology 551
Stratified ore deposits, econom. geol. 553
Stratigrafic geology 551
Strawberries agriculture 634
Street cleaning 628
railways engineering 625
Streets " 625
laying out of, hyg. of ground 614
local administration 352
trees in sanitary engineer'g 628
Strength of materials engineering 620
molec. phys. 539
Stresses bridge engineering 624
Strikes political economy 331
Stringd instruments 787
Strofe English prosody 426
Structural botany 581
geology 551
Structure, molecular physics 539
of the universe astron. 523
Strychnin materia medica 615
Stuarts English history 942.06
Stucco, external masonry 693
Student life and customs 371
Students' lives 923
Study furniture 651
rooms library economy 022
self-culture 374
of special topic see subject
Stuttering 616
Style fashion 390
literary rhetoric 808
Subject bibliografies 016
catalogs bibliografy 017
library admin. 025
Subjectivity knowledge 121
Sublime and beautiful esthetics 701
Submarine armor 627
cables telegrafy 654
mines military eng. 623
telegraf sociology 384
useful arts 654
tunnels mining eng. 622
warfare naval " 623
Submerged vessels ship-building 699
Subscription libraries 027
to creeds, eccles. polity 262
39 articles, Ang. ch. 283
Subsidies protection 337
Substance metaphysics 111
Subterranean telegraf 654
Subways hygiene of ground 614
road engineering 625
sanitation 628
Success in life ethics 174
Succession, apostolic eccles. pol. 262
monarchic, pol. science 321
to Eng. crown, const. law 342
property law 347
Suction pump, physics 533
Suez canal engineering 626
sociology 386
Suffixes English etymology 422
Suffocation protection of life 614.8
Suffrage 324
Sugar adulterations 614.3
cane agriculture 633
manufacture 664
organic chemistry 547
refinery 664
refining air pollution 614
Suicide customs 394
ethics 179
Suits in equity law 347
Sulfates chemical technology 661
Sulfur chemistry 546
compounds air pollu. 614
waters econ. geol. 553
Sulfuric acid, chemical technology 661
Summer-houses arbors 717
resorts amusements 791
schools self-education 374
Sumptuary laws ethics 177
Sun descriptiv astronomy 523

Topics in black-faced type are subdivided in tables.

Sun motors engineering 621
pictures photografs 770
worship 290
Sunda geology 559.2
history 992
religious history 279.2
travel 919.2
Sunday 263
closing library economy 024
school libraries 027
schools 268
Superintendence, building 692
school 379
Superior, Lake travel 917.7
Supernatural metaphysics 125
theology 231
Supernaturalism delusions 133
Superstitions 133
Supervision building 692
school 379
Supply and demand political econ. 338
Supremacy, papal, ecclesiastical polity 262
Supreme court reports, U. S. 345
Surface features of the earth 551
Surgeons' lives 926
Surgery 617
Surnames genealogy 929
Surveying geodesy 526
Surveys, general scientific 508.3
sanitary public health 614
Survival of the fittest evolution 575
Susceptibility, emotions 157
Suspension bridges 624
Svealand, Sweden history 948.7
travel 914.87
Swallows zoology 598
Swearing ethics 179
Sweating system laboring classes 331
Sweden antiquities 913.4
botany 581.9
geology 554.8
history 948.5
religious history 274.8
travel 914.85
zoology 591.9
Swedenborgian church 289
Swedish church sects 284
language 439
literature 839
Swimming 796
animal locomotion 591
Swindling confidence practices 174
Swine, diseases of comp. medicin 619
domestic animals 636
Switzerland antiquities 913.4
Switzerland botany 581.9
geology 554.9
history 949.4
religious history 274.9
travel 914.94
zoology 591.9
Sword exercise 796
Syenite lithology 552
Syfon physics 532
Syllogism logic 166
Symbolic logic 164
Symbolism Bible 220.6
religious art 246
Symfony orchestral music 785
Sympathy emotions 157
Symptoms diagnosis 616
Synagogs architecture 726
Synchronology history 902
Synod ecclesiastical polity 262
Synonyms English language 424
French " 444
German " 434
Greek " 484
Italian " 454
Latin " 474
Spanish " 464
Syntax comparativ philology 415
English language 425
Synthesis chemistry 545
logic 160
Syphilis diseases 616
public health 614
Syphon physics 532
Syria ancient history 939
Syriac language 492
Syrian church, sects 281
Syrup adulterations 614.3
Systematic botany 580
catalogs, bibliografy 017
theology 230
Systems manuals see subject
Tabernacle Biblical antiquities 220.9
sacred furniture 247
Table and parlor games 793
talk English literature 828
tipping spiritualism 133
Tableaux 793
Tables astronomy 524
furniture 642
library 022
interest 658
life life insurance 368
mathematical 510
mercantil 658
tide 525

Topics in black-faced type are subdivided in tables.

Tablets, memorial monuments 718
Tachygrafy shorthand 653
Tact mental faculties 151
Tactics, military 355
Tailoring 687
Takigrafy 653
Talbotypes 772
Talent mental faculties 151
Tales English literature 823
Talipes surgery 617
Talismans 133
Talmud Judaism 296
Tambourine instruments 789
Tamil language, Dravidian 494
literature 894
Tammany political associations 363
Tanning 675
air pollution 614
Taouism 299
Tapestry 746
Tapioca chem. technology 664
Tar " " 668
roofing building 695
Target-shooting 799
Targuins Chaldee versions 221
Tarifs 337
Tartary history 950
travel 915
Tasmania history 994
Taste esthetics 701
mental perceptions 152
Taverns 647
Tax titles law 347
Taxation local government 352
political econ. 336
Taxes, exemption from 336
tarif 337
Taxidermy 579
Tea adulterations 614.3
cultivation, agricult. 633
domestic economy 641
Teachers and teaching 371
institutes education 370
lives 923
Teaching of special topics see subject
Teas, commercial mixing of 663
Technical dictionaries 603
on special topics see subject
education 607
on special topics see subject
Technological schools 607
on special topics see subject
Technology 600
chemical 660
Teeth animals 591

Teeth dentistry 617
Teetotalism temperance 178
Tehuantepec ship canal engineer. 626
Telefone communication 654
electricity 537
sociology 384
Telegraf communication 654
electricity 537
military signals 623
sociology 384
Telegrafy useful arts 654
Teleology metaphysics 124
Telescope astronomy 522
optics 535
Telescopes famous 522
Temper ethics 179
Temperament piano 786
Temperaments mind and body 137
Temperance 178
hygiene 613
Temperature animal heat 612
conditions of life 577
hygiene 613
in diseases, diagnosis 616
measurement 536
of air meteorology 551
body physiology 612
earth geology 551
sense physiology 612
therapeutics 615
Tempering manufactures 671
molecular physics 539
Templars 366
knights 929.7
Temples architecture 726
student society halls 371
Temporal power papacy 262
Temptation doctrinal theology 233
Tenants law 347
Tender, legal paper money 332
Tenement houses architecture 728
Tenements hygiene 613
Teneriffe description 914.6
Tennessee, U. S. history 976.8
travel 917.68
Tension strength of materials 620
Tent life hygiene 613
Tenting outdoor sports 796
Tents as dwellings prehistoric arch. 571
Tenure of land law 347
political economy 333
office, civil service 351
library economy 023
teachers' 371
Teratology animals 591

Topics in black-faced type are subdivided in tables.

Teratology	natural hist. of man	573
	plants	581
	surgery	617
Terminology	metaphysics	112
Terms	schools	371
Terra cotta	art	738
	construction · masonry	693
	manufactures	666
Terra del Fuego	travel	918.2
Terraces	physical geology	551
Terrapin	zoology	598.1
Terrestrial magnetism		538
Tertiary age	geology	551
Test laws, Eng. religious history		274.2
Testacea		594
Testament, New		**225**
Old		**221**
Testamentary law		347
Testimony, legal	evidence	347
	probabilities	519
Tests	chemical analysis	543
Testudinata	zoology	598.1
Tetanus	diseases	616
	surgery	617
Teutonic knights		929.7
	minor languages	439
	literatures	839
	mythology	293
Texas, U. S.	history	976.4
	travel	917.64
	annexation of U. S. history	973.5
Text b'ks for writ'g verse, Eng. prosody		426
	instruction	371
Textil fabrics, inspection of		614.3
	manufactures	677
	prehistoric archeology	571
hangings	building	698
roofing	"	695
Texts	comparativ philology	418
	for learning Eng. lang.	428
	French "	448
	German "	438
	Greek "	488
	Italian "	458
	Latin "	478
	Spanish "	468
Thallofyta		589
Thanksgiving day	customs	394
	sermons	252
Thatch	roofing	695
Theaters	amusements	792
	ethics	175
	hygiene	613
Theatricals, private		793
Theft	law	343

Theism	natural theology	211
Theodicy	God	231
	providence	214
Theofanies		231
Theologians' lives		922
Theological education	religion	207
	seminaries	207
Theology		**200**
	biografy of	922
	natural	**210**
Theoretic astronomy		521
	chemistry	541
	ethics	171
Theories of construction	building	690
	special subject see subj.	
Theosofy		212
Therapeutics		615
Thermal springs	seismology	551
Thermochemistry		541
Thermodynamics		536
Thermometers	heat	536
	mechanic trades	681
Thermometry, meteorology		551
Thermoscope		536
Thessalonians	Bible	227
Thibet, China	history	951
Thirty-nine articles	creeds	238
Thirty years' war	European history	940
	German "	943
	Scandinavian hist.	948
Thomas à Kempis		242
Thomsonianism	cures	615
Thoracostraca		595
Thorough bass	music	781
Thought faculty		153
	transference	134
Thrift, laboring classes	polit. econ.	331
Throndjhem, Norway	history	948.4
	travel	914.84
Thugs		299
Thunder		537
	storms meteorology	551
Tide tables		525
	waters law	347
Tides	astronomy	525
	physical geografy	551
Tigers	zoology	599
Tiles	architectural construction	721
	building material	691
	manufactures	666
Tiling	building	695
Till	physical geology	551
Timber	architecture	724
	building material	691
	forestry	634

Timber manufacture 674
trade 338
Timbuctoo travel 916.6
Time chronology 529
cosmic 529
distribution of 529
keepers watchmaking 681
metaphysics 115
standard 529
tables, railroad 656
Timothy Bible 227
Tin chemistry 546
manufacture 671
metallurgy 669
mining 622
prehistoric 571
ores econ. geol. 553
roofing building 695
Tissues animal histology 591
vegetable " 581
Tithes 254
taxation 336
Title catalog library administration 025
to property, law of 347
Titles and duties library economy 023
changed book, bibliografy 014
of honor 929.7
Titus Bible 227
Toads 597
Toadstools botany 589
Tobacco, cerebro-cardiac poisons 615
cultivation 633
ethics 178
hygiene 613
inspection of adulterations 614.3
internal revenue 336
manufacture 679
state monopoly 336
Tobit Apocryfa 229
Toes, extra deformities 617
Toilet articles, customs 391
domestic economy 646
Tokio history 952
travel 915.2
Toleration, religious state ethics 172
Tolls, road and bridge 352
Tomb-burial public health 614
Tombs cemeteries 718
disposal of dead 393
Tongue diagnosis pathology 616
Tonic sol-fa vocal music 784
Tonics aids to digestion 615
Tonkin history 959
Tonnage gaging 389
shipping 656
Tools building 690
mechanic trades 680
mechanical eng. 621
prehist. archeol. 571
Toothache dentistry 617
Topical method instruction 371
Topografy hygiene of ground 614
maps 912
military, mil. science 355
surveying 526
Tories political parties 329
Tornadoes meteorology 551
Torpedoes explosivs 662
naval engineering 623
Torsion molecular physics 539
Tortoises zoology 598.1
Torts law 347
Torture customs of war 399
law 340
persecutions 272
Total abstinence 178
depravity doctrinal theology 233
Touch magnetism 538
mental faculties 152
physiology 612
Tournaments customs 394
Town halls architecture 725
officers local gov't 352
Towns government 352
sanitation of 628
Toxicology therapeutics 615
Toys amusements 793
Tract societies 206
Tractarianism Anglican church 283
Traction engines 621.1
Tracts, religious, practical 240
on special topics see topic
Trade, boards of commerce 381
depressions 381
domestic 381
foreign 382
free 337
list of books, publishers' 015
marks copyright 655
law 347
patents 608
slave 326
winds meteorology 551
Trades **mechanic** **680**
unions polit. econ. 331
useful arts **680**
Traditions folk-lore 398
Traffic commerce 656
Tragedies English literature 822
Training, physical hygiene 613.7

Topics in black faced type are subdivided in tables.

Training, physical outdoor sports 796
school hygiene 371
teachers 371
special, see subject
Trains railroad engineering 625
Traitors' lives 923
Tramps pauperism 339
Tramways city transit 388
engineering 625
mining engineering 622
Transactions gen. soc. 060
special, see subject
Transcendentalism 141
Transit instruments astronomy 522
Transits " 523
Translating, art of 808
Translations are clast with originals
principle of, comp. phil. 418
Transliteration comparativ philol. 411
in cataloging 025
Transmigration of souls 129
Transmission, heredity evolution 575
hygiene 613
machinery of eng. 6 1
of disease, pub. health 614
electric power, eng. 621.3
force physics 531
gases " 533
heat 536
power, mech. eng. 621
Transmutation of energy physics 531
metals chemistry 540
species, evolution 575
Transplanting, agriculture 634
Transportation, buildings arch. 725
criminal law 343
military 355
sociology 380
useful arts 656
Transubstantiation 265
Transvaal, Africa history 968
Transversals modern geometry 513
Transylvania history 943.9
Trap lithology 552
Trapping business 639
Trappists monastic orders 271
Traumatic fever surgery 617
Travel hygiene 613
Travelers, manuals for 910
protection of, pub. health 614.8
Travelers' conversation books, Eng. 428
lives 923
Traveling, art of 910
Travels, general 910
scientific 508.3
Travels special see subject
Traverse tables surveying 526
Treadmills mechanical eng. 621
punishments 343
Treason law 343
Treasury notes paper money 332
U. S., administration 353
Treaties international law 341
Treating temperance ethics 178
Treatises, Am. and Eng. priv. law 347
on special topics, see subject
Treatment of captivs war customs 399
disease 616
the dead, customs 393
pub. health 614
Treaty of Washington 341
Trees botany 582
forestry 634
fruit culture 634
in streets, sanitation 628
ornamental 715
Trefining surgery 617
Trent creed of Pius 4 238
Trial civil 347
court-martial 344
criminal 343
ecclesiastical 262
jury 340
ordeal 340
state 343
Trials, civil law 347
criminal " 343
military " 344
Triangle musical instruments 789
Triassic age geology 551
Tribes form of state 321
Trichinae economic zoology 591
parasitic diseases 616
Trichiniasis public health 614
Tricks with cards amusements 793
Tricycling 796
Triest history 943.6
travel 914.36
Trigonometry 514
Trilobites paleontology 565
zoology 595
Trinity doctrinal theology 231
Tripoli, Africa history 961
"Triumfs" student customs 371
Troas ancient history 939
Trombone wind instruments 788
Tromso, Norway history 948.4
travel 914.84
Tropical plants 581.9
Trotting racing 798

Topics in black-faced type are subdivided in tables.

Troubadours 849
Trout culture, fishing 639
Trouvères French literature 841
Troy ancient history 939
Truces international law 341
Truck system labor and wages 331
Trumpet wind instruments 788
Trunkmaking 685
Trusses engineering 624
surgery 617
Trust companies banks 332
Trusts and trustees **law** 347
Truth ethics 177
Tubes, pneumatic communication 654
speaking " 654
Tubular bridges 624
Tudor architecture 724
Tudors English **history** 942.05
Tungusic language 494
Tunicata 594
Tunis history 961
travel 916.1
Tunneling mining engineering 622
Tunnels aqueducts 628
Turanian languages 494
literatures 894
Turkestan geology 555.8
history 958
religious history 275.8
travel 915.8
Turkey in Asia geology 555.6
history 956
religious history 275.6
travel 915.6
Europe botany 581.9
history 949.6
travel 914.96
Turkish architecture 723
baths customs 391
hygiene 613
harems customs 392
language 494
literature 894
Turning lathes 621
Turnpikes road **engineering** 625
Turtles **zoology** 598.1
Twilight 525
Tyfoid fever pathology 616
public health 614
Tyfus fever pathology 616
public health 614
Tyler, J: presidency of 973.5
Type and type setting printing 655
founding 655
writers communication 652
Typografers' lives 926
Typografy 655
Typology 220.6
Tyrian architecture 722
Tyrol history 943.6
travel 914.36
Tyrone, Ireland history 941.6
Tyrotoxicon animal poisons 615
Ulcers surgery 617
Ulster, Ireland history 941.6
travel 914.16
Ultramontanism catholic church 282
eccles. polity 262
Unconsciousness metaphysics 127
Unction, extreme theology 265
Underground roads engineering 625
mining eng. 622
Understanding mental faculty 153
Undulation dynamics 531
sound 534
Undulatory theory of light 535
Uniformity, act of, Eng. religious hist. 274.2
Uniforms, military 355
Unionidae zoology 594
Unions, trade polit. economy 331
Unique books, rarities 094
Unitarian church 288
Unitarians' lives 922
United Americans political soc. 363
brethren sects 284
Kingdom **history** 942
Netherlands " 949.2
presbyterian church sects 285
Provinces history 949.2
service, military and naval 355
States administration 353
antiquities 913.7
bank 332
botany 581.9
boundaries 973
Christian com. relig. 267
U. S. hist. 973.7
coast survey 526
constitutional law 342
finance 336
geology 557.3
government admin. 353
history 973
laws 345
literature 810
religious history 277.3
sanitary com., charity 361
U. S. hist. 973.7
statistics 317
statutes 345

Topics in black-faced type are subdivided in tables.

United States travel 917.3
zoology 591.9
United States of Colombia geology 558.6
history 986
relig. hist. 278.6
travel 918.6
Unity metaphysics 111
of God doctrin 231
human race 572
knowledge 121
language 401
nature 501
Univalves mollusks 594
Universal gravitation, law of mechan. 531
history 909
language 408
suffrage 324
Universalism 289
Universalists' lives 922
Universals metaphysics 113
Universe astronomy 523
Universities 378
architecture 727
University education 378
extension 378
of the State of New York 379
people's, library as the 021
Universology cosmology 113
Unregenerate doctrinal theology 233
Upholstering furniture 684
Ural-Altaic language 494
literature 894
Uranografy astronomy 523
Uranus planets 523
Urchins, sea radiates 593
Uruguay geology 558.9
history 989
religious history 278.9
travel 918.9
Usages, social, etiquet 395
Use of words Eng. philology 428
Useful arts 600
bibliografy of 016.6
biografy of 926
Usefulness of animals, econ. zool. 591
plants " bot. 581
Usury banking 332
law 347
Utah, U. S. history 979.2
travel 917.92
Utilitarianism ethics 171
Utility of philosofy 101
special subjects, see subject
Utopias political science 320
Vaccination cont. diseases 614

Vagabonds pauperism 339
Valleys physical geografy 551
Value, standards of, political economy 332
Values, average probabilities 519
Valves hydraulics 532
mechanical eng. 621
Vampire zoology 599
Van Buren, M., presidency of 973.5
Vancouver's island history 971
Vapor engines mech. eng. 621
Vaporization heat 536
Vapors meteorology 551
properties of 533
Variation evolution 575
in animals, phys. zool. 591
of compass 522
Variations, calculus of 517
Variolation contagious diseases 614
Varioloid diseases 616
Varnishes chemical technology 667
Varnishing painting 698
Varnishmaking air pollution 614
Vases pottery 738
Vassalage 321
Vassar college, colleges for women 376
Vatican council, 1870 infallibility 262
Vaticination prediction 133
Vaudois Christian sects 284
persecution 272
Vaults air pollution 614
arch. construction 721
design 729
Vedas Sanskrit literature 891
Vegetable anatomy 581
chemistry 547
diet hygiene 613
fats chem. tech. 665
food hygiene 613
irritants poisons 615
oils chem. tech. 665
parasites, diseases 616
physiology 581
substances, analysis of 543
trades air pollution 614
world botany 580
Vegetables, agriculture 635
inspection of, pub. health 614.3
preservation of, chem. tech. 664
Vegetarianism hygiene 613
Vegetation botany 580
Vehmgericht criminal law 343
Veins, geology 551
mineral economic geology 553
Vellum, books printed on rarities 096
Velocipede 796

Topics in black-faced type are subdivided in tables.

Venders	law	347
Veneering	architectural decor.	729
Venesection	therapeutics	615
Venezuela	geology	558.7
	history	987
	religious history	278.7
	travel	918.7
Ventilating	science	536
Ventilation	building	697
	library economy	022
	mining engineering	622
	sanitary "	628
Ventriloquism		791
Venus	planets	523
	transits of astron.	523
Vermes	articulates	595
Vermont, U. S.	history	974.3
	travel	917.43
Verse, text-books for writing		426
Versification, English prosody		426
Vertebrates	**paleontology**	**566**
	zoology	**596**
Vespers	music	783
	ritual	264
Vestments		247
Vesuvius	volcanos	551
Veterinary medicin		619
	sanitation, pub. health	614
Viaducts	bridges	624
Viaticum	sacrament	265
Vice	crime	343
	ethics	170
Victoria, Queen	English history	942.08
Australia	"	994
Vignettes	**engravings**	**760**
Vikings	history	948
Village communities		321
Villas	architecture	728
Villeins	feudalism	321
Vine culture	agriculture	634
Vinegar		664
Vines	botany	581
Vineyards		634
Violins	stringd instruments	787
Virgin Mary	catholic doctrin	230
Virginia, U. S.	history	975.5
	travel	917.55
West, U. S.	history	975.4
	travel	917.54
Virginity	religious asceticism	248
Virtues		**170**
Visible speech, phonology		414
Visigoths	German history	943
Vision, disorders of, asylums for blind		362
	diseases of eye	617
Vision	mental faculty	152
	optics	535
Visions and dreams		135
Visitation and search	internat. law	341
	pastoral	253
Visitations	English heraldry	929.7
Vital force	biology	577
functions	animal physiol.	591
	human "	612
phenomena, biology		577
statistics	population	312
	public health	614
Vitalism	biology	577
Vitality	zoology	591
Vivisection	animal anatomy	591
	ethics	179
	laws, public health	614
Vocal culture	elocution	808
	music	784
hygiene, etc.		784
music		784
organs, acoustics		534
	human anatomy	611
	physiology	612
Vocalists' lives, singers		927
Voice	elocution	808
	physiology	612
	singing	784
Volapük	universal language	408
Volcanic rocks	lithology	552
Volcanos	physical geology	551
Volkslied	vocal music	784
Voltaic electricity	physics	537
Volumetric analysis	chemistry	545
	geometry	513
Voluntary	music	786
Volunteers	army	355
Vomiting	diseases	616
Voodooism		133
Vortex rings	molecular physics	539
Voters, qualifications		324
Voting	suffrage	324
machines	suffrage	324
Vowels	English language	421
Vows	asceticism	248
Voyages, Arctic		919.8
round the world		910
scientific		508.3
travels	adventures	910
Vulgarisms	English language	428
Vulgate	Bible	220.4
Wager of battle	legal antiquities	340
Wagers	ethics	175
Wages, labor and, polit. economy		331
Wagnerian music		782

Topics in black-faced type are subdivided in tables.

Wagons	manufacture	684
Wake	funeral customs	393
Wakefulness	insomnia	613
Waldenses	persecutions	272
	sects	284
Wales	geology	554.2
	history	942.9
	language	491
	religious history	274.2
	travel	914.29
Walking	animal locomotion	591
	matches amusements	796
	pedestrianism	613.7
Walks	landscape gardening	713
Wall papers	decoration	745
	manufacture	676
	paper hanging	698
street	banking	332
Wallachia	history	949.8
	travel	914.98
Wallachian language		459
literature		859
Walls arch. construction		721
design		729
Waltz	music	786
Waltzing	amusements	793
	ethics	175
Wandering Jew	legends	398
War captures	law	341
cause of	"	341
civil	U. S. history	973.7
customs of		399
dances		399
department, U. S.		353
ethics		172
French and Indian, U. S. history		973.2
law of		341
Mexican	U. S. history	973.6
of 1812,	" "	973.5
Roses	English "	942.04
Revolutionary	U. S. history	973.3
science and art of		355
woman in	woman's position	396
Warehouses	architecture	725
Warm baths	hygiene	613
Warming	building	697
	sanitary engineering	628
Warnings	omens	133
Warpt surfaces	descr. geometry	515
Warrants	law	343
Warts	skin diseases	616
Washing	laundry	648
Washington, G: presidency of		973.4
D. C.	history	975.3
	travel	917.53
Washington, U. S.	history	979.7
	travel	917.97
treaty of		341
Wasps	insects	595
Wastes, disposal of	sanitation	628
Watches	horology	529
Watchmaking		681
Water, artificial ponds, etc.		714
as a beverage	hygiene	613
chemical analysis		543
closets	house drainage	628
colors		751
cure	therapeutics	615
engines	mechanical eng.	621
filters	hygiene	613
force of	engineering	627
gas	heating	697
glass	chem. tech.	666
hydraulics		532
microscopic analysis, hygiene		613
mineral, artificial		663
physical geografy		551
pollution		628
power	mechanical eng.	621
products		338
supply, canal engineering		626
	sanitary "	628
towers	fire extinction	614.8
travel	protection of life	614.8
wheels	hydraulics	532
	mech. engineering	621
works		628
	city administration	352
Watering places	hygiene	613
Waterloo, battle of	French history	944
Waters, mineral	adulterations	614.3
	geology	553
	hygiene	613
	therapeutics	615
Wave theory	optics	535
Waves, theory of	hydraulics	532
Wax flowers		746
Wealth and rank, social ethics		177
political econ.		331
Weapons	customs	399
	military science	355
	prehist. arch.	571
Weather	meteorology	551
Weaving		677
Wedding customs		392
Wedgewood ware	pottery	738
Weeds	economic botany	581
	pests	632
Week, names of days of		529
Weems	prehistoric arch.	571

Topics in black-faced type are subdivided in tables.

Weighing and measuring chem. 542
Weights and measures 389
Well water impurities hygiene 613
Wellesley college, colleges for women 376
Wells physical geografy 551
water supply 628
Welsh history 942.9
language 491
literature 891
travel 914.29
Were-wolves folk-lore 398
West Indies geology 557.2
history 972.9
travel 917.29
Point academy 355
U. S. travel 917.8
Virginia, U. S. history 975.4
travel 917.54
Western or Rom. catholic church 282
states, U. S. mountain
geology 557.8
history 978
relig. hist. 277.8
travel 917.8
Pacific geology 557.9
history 979
relig. hist. 277.9
travel 917.9
Westminster assembly's catechism 238
confession creeds 238
Whale fisheries 639
zoology 599
Wheat adulterations 614.3
agriculture 633
trade, agricul. products 338
Wheel and axle, physics 531
cycling amusements 796
Wheels, cog mechanical eng. 621
railroad engineering 625
water hydraulics 532
mechanical eng. 621
wooden, manufactures 674
Whigs Eng. political parties 329
U. S. " " 329
Whirlpools ocean currents 551
Whirlwinds meteorology 551
Whiskey chemical technology 663
insurrection, Pa., hist. 974.8
tax 336
Whist games 795
White hair dermatology 616
lead manufacture 667
Whooping cough pathology 616
public health 614
Wife's duties ethics 173
Wigs customs 391
Will metaphysics 159
of God ethics 171
theology 234
William 1, Conqueror, Eng. history 942.02
2 " " 942.02
3 " " 942.06
4 " " 942.07
Wills genealogy 929
law 347
Wilmot proviso U. S. history 973.6
Wind and current charts 551
instruments music 788
mills air engines 621
Window gardens 716
Windows, arch. construction 721
design 729
staind glass arch. decor. 729
Winds meteorology 551
Wine adulterations 614.3
at table temperance 178
domestic economy 641
manufacture 663
temperance 178
Wings animal locomotion 591
Wire manufacture 671
Wisconsin, U. S. history 977.5
travel 917.75
Wisdom Apocryfa 229
divine, doct. theology 231
Wit American humor 817
English " 827
French " 847
German " 837
Greek " 887
Italian " 857
Latin " 877
Spanish " 867
Witchcraft 133
Witches persecutions 272
Witnesses evidence 347
Wives' duties family ethics 173
Woman customs 396
education 376
suffrage 324
political parties 329
woman's position 396
Woman's position and treatment 396
rights woman's position 396
Women, diseases of 618
duties of family ethics 173
social " 177
labor of political econ. 331
lives of collections 920.7
Women's religious societies 267

Topics in black-faced type are subdivided in tables.

Women's rights 396
 work in church 259
Wood building material 691
 carving 736
 construction, general, carpent. 694
 engraving 761
 implements manufactures 674
 working " 674
Woodbury types 773
Woods economic botany 581
 forestry 634
Wool, agriculture 636
 dyeing chemical technology 667
 manufactures 677
Woolen clothing hygiene 613
Word methods, elemen. education 372
Words English etymology 422
 slang 427
 use and abuse, Eng. 428
 vulgarisms " 428
Workhouses, asylums 362
 parish, societies for 256
 Sunday 263
Working classes political econ. 331
Workshop receipts, chem. tech. **660**
World, atlases of 912
 voyages round 910
World's fairs 606
 plurality of 523
Worms zoology 595
Worship family 249
 of animals idolatry 290
 public 264
Worsteds manufactures 677
Wounded, hospitals for 362
Wounds surgery 617
Wrecking and life-saving 614.8
Wrestling 796
Writing 652
 books 372
 for the press lit. methods 029
 inks chemical technology 667
 machine 652
 materials 652
 of history 907
 shorthand 653
Writs law 347
Written examinations education 371
Wrongs torts law 347
Wyoming, U. S. history 978.7
 travel 917.87
 massacre Penn. history 974.8
Xylofone musical instr. 789
Xylografy 761

Yacht-building 699
Yachting 797
Yarns manufactures 677
Yearbooks statistics **310**
Year books, see subject
Yeast adulterations 614.3
 botany 589
 chemical technology 664
 domestic economy 641
Yellow fever diseases 616
 public health 614
Yellowstone park description 917.87
York, house of English history 942.04
Yosemite, Cal. description 917.94
Young, counsels to ethics 173
 education of elemen. educa. 372
 men, advice to, business ethics 174
Y. M. C. A. 267
 buildings arch. 726
Young men's Christian unions 267
 people's religious societies 267
 reading of 028
 women, advice to, social ethics 177
 women's Christian ass'n 267
Yucatan, Mexico architecture 722
 history 972
 travel 917.2
Zambezi travel 916.7
Zanzibar, Africa history 967
Zechariah Bible 224
Zend Avesta Persian lit. 891
 religion 295
 language 491
Zephaniah Bible 224
Zinc inorganic chemistry 546
 metallurgy 669
 ores economic geology 553
 roofs building 695
Zither music 787
Zodiac astronomy 523
Zodiacal light " 523
Zoofytes 593
Zoological gardens 590
Zoology **590**
Zoosporeae botany 589
Zoroaster Parseeism 295
Zouaves military science 355
Zululand history 968
Zunis Indians 970.3
Zurich history 949.4
 travel 914
Zwinglian Protestant sects 284
Zygosporeae algae 589
 lichens 589

Topics in black-faced type are subdivided in tables.

SUBJECT INDEX SUPPLEMENT

Abattoirs	architecture	725
Abbreviations	English paleografy	421
	Greek paleografy	481
	Latin paleografy	471
Abchasian language		494
Abecedarians	christian sect	284
Abiogenesis	animals	591
	biology	576
	plants	581
Absorption	physiology	612
Academic education		373
Academies, private	for boys	373
	girls	376
	mixt	373
Accidents, railroad		656
Acclimation	evolution	575
	mental influences	136
Accounts, household		647
Acetylene	chemic technology	665
Acetylenes	organic chemistry	547
Acrobats	ethics	175
Acrogens	botany	586
Adamites	heresy	273
Adaptation	evolution	575
Adding machines		510
Administration, household		647
Adoptionists	heresy	273
Adsorption	pneumatics	533
Aerial navigation	engineering	629
Aeronautics	"	629
Aeroplanes	"	629
African languages, Ethiopic		492
Agricultural colony	labor'g classes	331
Air, liquid	physics	533
Aladdin oven	dom. econ.	643
Alarm, fire	local gov't	352
Alcohol, denatured		661
Aldehydes	organic chemistry	547
Alfalfa	agriculture	633
Algrafy	lithografy	763
Alliteration	prosody	416
	rhetoric	808
Altars	architecture	729
Ambidexterity		612
Anglican chant		783
sects		283
Anglo-Saxon race characteristics		136
Animals	zoology	590

Anniversaries, U.S. hist.	civil war	973.7
	revolution	973 3
Annotations for readers		028
Antarctic regions	travel	919.9
Antirust processes	bldg	691
Apetalae	botany	583
Appendicitis	diseases	616
	surgery	617
Appraisal of books		028
Arabesques	decoration	729
	design	745
Arctic travels		919.8
Argumentation	rhetoric	808
Aroostook war	U. S. history	973.5
Arrest	civil trials	347
	criminal law	343
Arsenic ores		553
Arts and crafts		740
Asbestos roofing		695
Asparagus	agriculture	635
Asphalt	chemic technology	665
	economic geology	553
	pavements	625
	roofing	695
Astringents	therapeutics	615
Astrophysics	astronomy	523
Atavism	evolution	575
Atlantic telegraf	elec. engin'g	621.3
Auditing	accounts	657
Audubon societies		179
Augury		133
Auroras	astronomy	523
Austral islands		996
Authority, school		371
Automobiles	engineering	629
Automobiling	amusements	798
Autonomy	ethics	171
	polit. science	320
Avars	history	950
Avatars	Brahmanism	294
Aviation	engineering	629
Babism	Mohammedan sect	297
Bacon's rebellion	Virginia history	975.5
Badminton	games	796
Bags and bagging		677
Ballast	transportation	656
Balloons, dirigible	engineering	629
Basket ball	games	796

Topics in black faced type are subdivided in tables.

Basketry	manufactures	674
	ornamental	745
	prehistoric	571
Beads	decoration	746
	ornaments	391
Beadwork	decoration	746
Beans	agriculture	635
Bedbugs	pests	648
	zoology	595
Bees	agriculture	638
Beetles	pests	632
Bering sea question	U. S. hist.	973.8
Berries	fruits	634
Bertillon system	anthropology	573
Bibliofily		010
Biennial legislation		328
Birds	agric. aids and pests	632
	diseases of	619
	game sports	799
	poultry	636
	protection	179
Births	registration	614
Bitters	therapeutics	615
Bleeding	surgery	617
	therapeutics	615
Bloody assizes	English history	942.06
Blowpiping	quantitativ analysis	545
Boats, life	lifesaving service	656
	public safety	614.8
	shipbuilding	623
Bogies	folklore	398
Bolak	universal language	408
Boss	Amer. politics	329
Bowling	amusements	796
Boxer uprising	Chinese history	951
Brownian motion	physics	539
Bubonic plague	public health	614
Bulbs	horticulture	716
Butchershops		641
Cables	electric engineering	621.3
Cadastration	taxation	336
Calcimining		698
Calibration	instrument making	681
Calorimetry	physics	536
Calotype	photografy	772
Canals	transportation	656
Canonization	dogma	235
Capitalization	English language	421
	securities	332
Capon	poultry	636
Care of children		649
Carnegie institution		378
	pension fund	371
Carriages, motor	engineering	629
Carriers, mechanical book		022
Cars, motor	engineering	629
Casein	chemic technology	668
Cases, book	library economy	022
Catabaptists	sects	284
Caterpillars	agriculture	632
	economic zoology	591
	silk culture	638
	systematic zoology	595
Causality	philosofy	122
Chairs	domestic economy	645
Characteristics, mental		136
Charismata	theology	231
Charms	occultism	133
Charts of statistics		311
Chemistry and physics, interrelations		541
Chicken farming		636
Childstudy		136
Chills and fever	diseases	616
	pub. health	614
Chimney money	taxation	336
China ware	domestic economy	642
Chinook language		427
Chirognomy	palmistry	133
Chiropody	surgery	617
Christian science healing	law	614
	sect	289
	medicin	615
	theology	230
Chromophotografy		770
Chromoxylografy		761
Church biografy		922
Church property	mon. foundations	271
	polit. science	322
	taxation	336
Cigar making		679
Cities	beautification	710
Citizenship	suffrage	324
Civil rights	constitutional law	342
	legal capacity	347
	U. S. history	973.8
Cleaning	chemic technology	667
	domestic economy	648
Clothes	" "	646
Collecting	business methods	658
Collodion process	photografy	772
Colophons	incunabula	093
Colors, painting	trade	698

Topics in black faced type are subdivided in tables.

Combinations, industrial
political economy 338
Combs customs 391
Commerce, interstate 385
Commissions, library 027
Communal marriage 392
Communication electric engin'g 621.3
Commutators electric engineering 621.3
physics 537
Complex ores metallurgy 669
Composit order of architecture 729
Compromise, Missouri U. S. hist. 973.5
of 1850 " " 973.7
Concrete chemic technology 666
Condensed milk public helth 614.3
Conduction electricity 537
heat 536
Conferences, library 020
Conics analytic geometry 516
modern geometry 513
Conscription U. S. history 973 7
Constitution of matter metaphysics 117
physics 530
Consumers leag 331
Consumption political economy 339
Continuation schools 374
Conveyers engineering 621
Cookery domestic economy 641
Cooks domestic economy 647
Copiers office equipment 652
Corea history 951
Cornish pumps mine drainage 622
Coroners law 347
Corpuscular theory of light 535
matter 530
Correlation method of teaching 371
Correspondence autobiografy 920
commercial 658
schools, self education 374
Corruption, official ethics 172
Cost keeping accounts 657
Country life 630
Courses curriculum 375
Cradle songs folklore 398
music 784
Creameries agriculture 637
Cries, battle war customs 399
street folklore 398
Criminology 364
Currents, electric physics 537
Dangers to life, protection against 614.8
Day nurseries 362
Decoration design 740
domestic economy 645
Demurrage transportation 656
Deontology 170
Design metaphysics 124
arguments from 211
Dictation machines 651
Didactics education 371
Direct legislation polit. science 321
primaries suffrage 324
Disarmament arbitration 341
Dishes table ware 642
Dompelaers baptist sect 286
Drafting drawing 744
Drafts games 794
Dragonade persecutions 272
Dragons folklore 398
Drawing, mathematical 744
Drawnwork fancywork 746
Dryplate processes photografy 772
Duplicating machines 652
Dynamic engineering 621
Dynamoelectric machinery 621 3
Ear, care of hygiene 613
Eastern church 281
Education, public system 379
El dorado folklore 398
Electric vehicles engineering 629
Electricity, applied " 621.3
Electroanalysis chemistry 544
Electrobiology biology 577
Electrophysiology biology 577
Electrons chemistry 541
Electrotherapy medicin 615
Engineering, dynamic 621
Engineering, electric 621.3
illuminating 628
industrial 658
English language 420
Esperanto 408
Ether light 535
physics 530
Eudemonism ethics 171
Eugenics evolution 575
Evaluation of books 028
Evil eye 133
Exotics horticulture 716
Facsimiles class with originals

Topics in black faced type are subdivided in tables.

Factory management		658
Fashions	costume	391
Favorits, royal	biografy	923
Fear	psychology	157
Feldspar	economic geology	553
	mineralogy	549
Finger prints	anthropology	573
Fireplaces, artistic		749
Fireproofing	building	693
First editions	rarities	094
Fish, diseases of		619
trade	economics	338
Fishways	fish culture	639
Fleur de lis	heraldry	929.6
Flowers, national		929.6
Flying machines	engineering	629
Fools	customs	394
Forage	agriculture	633
Forging	blacksmithing	682
Formaldehyde	chemistry	547
	public health	614
Frauds, art		709
literary		809
Freight	interstate commerce	385
Fylfot	ecclesiólogy	246
	heraldry	929.6
Galvanoplasty	**manufactures**	**671**
Game	sport	799
Garages, private	architecture	728
public	"	725
Gargoyles	architec. decoration	729
Gasoline engines		621
vehicles		629
Gelatin manufacture		668
Geomorfology	geology	551
Ginseng	agriculture	633
Gipsy moth	pests	632
Gouache painting		751
Graft	business ethics	174
	political "	172
Halftone process	photografy	777
Hall marks	plate	736
Halos	religious art	246
Hektograf	office appliances	652
Heliogravure	photografy	777
Hives	apiculture	638
Hocky	outdoor sports	796
Home	domestic economy	643
economics		**640**
House	domestic economy	643
care of	" "	648
tax	political "	336
Household administration		647

Humane societies		179
Ice sports		796
Ideal states	political science	320
Illuminating engineering		628
Imperialism	political science	325
Implements,	primitiv	571
Industries	"	571
Infants, diseases of		618
Initials	bibliografy	014
	decoration	745
Inlay	art furniture	749
Insect powders		648
Insecticides	agriculture	632
	domestic economy	648
Insectifuges	" "	648
Insignia	heraldry	929.6
Insurance, teachers		371
International relations	polit. sci.	327
Internationalism		341
Invalids, care of	domestic economy	649
Ions	chemistry	541
Iron work, ornamental		739
Ivory painting		751
Japanese color prints		761
Jumping	athletics	613.7
Juvenil courts		364
Kalendar		**529**
Kinetoscope	photografy	778
Kinship	customs	392
Kodaks	photografy	771
Koomiss, *see* Kumiss		
Koumiss, *see* Kumiss		
Krypton	inorganic chemistry	546
Kumiss	adulterations	614.3
	dairy products	637
Laborsavers, domestic		643
Legislativ manuals		328
Legumes	agriculture	635
Lenses	photografy	771
Libraries for blind		027
Lighting	engineering	628
systems	domestic econ.	644
Liquid air	physics	533
Love	psychology	157
Luster ware	pottery	738
Luxury	economics	339
	ethics	177
Lyddite	chemic technology	662
Madonna, legends of		225
Mary, Virgin	biografy	225
Menageries	amusements	791
Mending	domestic economy	646
Metalwork, artistic		739

Topics in black-faced type are subdivided in tables.

Milk	dietetics	613
Mimeograf	office fittings	652
Mind	childstudy	136
Molding	metal manufacture	671
Monsters, fabulous	folklore	398
Moth, gipsy	pests	632
Moths	agriculture	632
Motor vehicles		629
Mountaineering	hygiene	613.7
	sports	796
Mushrooms	culture	635
National conservation		333
resources	conservation	333
Nations, origin of		901
rise of	**Europ. history**	**940**
Natural law	science	501
Nature, law of	science	501
unity of	metaphysics	110
	science	501
Neostyle	office equipment	652
Nicaragua canal	engineering	626
	sociology	387
Noise	public health	614.8
Norwegian haybox	cookery	641
	utensils	643
Numbering machines	office	652
	lib. econ.	025
Nurses	domestic service	647
Nuts	agriculture	634
Obsession		133
Occultism		133
Od	mind and body	134
Odylic force		134
Ore sampling	assaying	669
Ores	chemic analysis	543
Organ	musical instruments	786
Organization	central gov't, U S	353
Osteopathy	therapeutics	615
Oxids	mineralogy	549
	theoretic chemistry	541
Oxygen compounds	mineralogy	549
Paleografy	inscriptions	417
Pamirs	history	958
Paneld construction	carpentry	694
Panels	arch. decoration	729
Panics, commercial	money	332
	production	338
	trade	381
Pantomimes	drama	792
Paralysis	diseases	616
infantil	diseases	616
	public health	614
	surgery	617
Parsifal	legends	398
	opera	782
Pasteurization	infant diet	613
Patent medicins	therapeutics	615
Pestifuges	domestic economy	648
Pests, household	" "	648
Philadelphia	exposition 1876	606
Photogammetry	photografy	778
Physical chemistry		541
Physicochemistry		541
Pianola		789
Picture frames	decoration	749
Pigeons	domestic birds	636
Pigments	fine arts	751
Pigmies	folklore	398
Plan of house	domestic economy	643
Political history		**900**
Polyglot dictionaries		413
Pragmatism	philosofy	149
Prevention of cruelty	ethics	179
Prisms	lighting	628
Probation of malefactors		364
Promiscuity	customs	392
Psalms	hymnology	245
Psychologic medicin	diseases	616
Public service commissions		380
Pump, air	practical chemistry	542
Pyrografy	fine arts	736
Raffia	agriculture	633
Rarities	**bibliografy**	**090**
Rat	disease carrier	614
	domestic economy	648
Receptions	" "	642
Refreshments	" "	642
Regattas	amusements	797
	college and school	371
Registration of land	Torrens system	333
Relations, international		327
of church & state	pol. sci.	322
Relief work	charity	361
	parish	258
Religious ceremonies	ritual	264
Rice	agriculture	633
Rolling mills	metal manufacture	671
Roman empire, Holy		943
Rope walking	entertainments	791
Rugs	domestic economy	645
	manufactures	677
	design	745
Sage foundation		306
Saints in art	ecclesiology	246
	painting	755

Topics in black-faced type are subdivided in tables.

Saints, Latter day	doctrin	298
Samaritans	sects	296
Sanitary measures	dom. econ.	648
Satanism	doctrinal theology	235
	witchcraft	133
Savoy, house of	Ital. history	945
Scholarships	research	378
	school tuition	371
Screens, food		648
Scriptural biografy	N. T.	225
	O. T.	221
Sectarian teaching in schools		377
Semitic literatures		892
Separators, cream	dairy	637
Sewing, plain	dom. economy	646
Shaving	customs	391
	toilet	646
Shelter	domestic economy	643
Shields	heraldry	929.6
	weapons	399
Shintoism		299
Silver question	political economy	332
Single tax		336
Skeeing	sports	796
Skiing	sports	796
Skin dressing	taxidermy	579
Slaughtering industry		664
Slavic literatures		891
Slide rule	math. instruments	510
Sliding shelvs	library storage	022
Sloyd	education	371
Smithsonian exchanges	lib. econ.	021
Snowshoeing	sports	796
Societies, college		371
secret		366
Somatology	anthropology	573
Sounds, systems of	phonology	414
Southwest, U S	history	978
Splügen pass	Alps	914.94
Stacks	library economy	022
Stadia	surveying	526
Stage coach	transportation	656
Stairs, moving	mechanical engin'g	621
Stamp collecting		383
State aid to libraries		021
and church	pol. science	322
Steam engineering		621.1
Stencil decoration		747
Sterilization	food preservation	641
	infant feeding	613
Still life	painting	754
Stores, department		658
Stoves	kitchen equipment	643
Streams, pollution	factory waste	628
Street sprinkling		628
Student self-government		371
Study clubs		374
Sudan, Africa	history	966.2
Egypt	"	962.6
Suggestion	hypnotism	134
Surinam, S. Amer.	history	988
Swastika	ecclesiology	246
	heraldry	929.6
Swedish gymnastics	hygiene	613.7
Swindling	law	343
Swords	war customs	399
Syria	modern history	956
Systems, philosofic		140
Table	domestic economy	642
Tables, furniture	" "	645
Taboo	religion	299
Talbotypes		772
Tar roads		625
Tattooing	customs	391
Tea	beverages	641
Telegraf codes		652
Telepathy	mind and body	134
Telpherage	mechanical engin'g	621
Telephot	photografy	778
Tenements	domestic economy	647
Tennis	lawn	796
	parlor	794
Terbium	inorganic chem.	546
Teutonic race	ethnology	572
Textbooks, free	public schools	379
state	" "	379
Thermometry	physics	536
	physiology	612
Tintypes	photografy	772
Tires, rubber	manufactures	678
Tobacco pipes	manufactures	679
Tobogganing	sports	796
Token money	numismatics	737
	political economy	332
Torrens system		333
Totemism		392
Trade, balance of	commerce	382
catalogs	bibliografy	015
Trading companies		338
stamps		658
Tramways	electric engineering	621.3
Trance	hypnotism	134
Traps	domestic economy	648
	sanitary engineering	628
Traveling libraries	lib. econ.	021
Tricks	sleight of hand	791

Trout fishing	sports	799
Trucks, book	library economy	022
Trustees, library		023
Trusts	corporations	338
Tuberculosis	diseases	616
	public health	614
Turpentine	chemic technology	665
Upholstery	domestic economy	645
Use of library	library economy	024
Vacation schools		374
Vacations, school		371
Vacuum cleaners	domestic econ.	648
Values, educational	curriculum	375
food	domestic economy	641
Vampire	folklore	398
Vapor lamps	electric lighting	621.3
Vaulting	athletics	613.7
Veneer	manufacture	674
Veneering	cabinet work	684
Ventilation	domestic economy	644
Vice	education	379
Villains	feudalism	321.3
Villenage	"	321.3
Virgin Mary	biografy	225
Vocational education		371
Voltaic cells	engineering	621.3
Voyages, Antarctic		919.9
War dances		399
Water marks	paper manuf.	676
power	hydraulic engin'g	627
underground	phys. geog.	551
Waterproofing	building	691
Wave theory	sound	534
Weirs	hydraulic engineering	627
	" machinery	621
	hydraulics	532
Western hemisphere	geografy	917
Wharves	engineering	627
Whipping post		343
Whisky traffic		178
Winter resorts	hygiene	613
	sports	796
Wire	manufacture	671
Wolfram	inorganic chemistry	546
Wood alcohol	chemic technology	661
Wringers, clothes	laundry	648
Xenon	inorganic chemistry	546
Young	child study	136
counsels to	ethics	173
Ytterbium	inorganic chemistry	546
Yttrium	" "	546
Zincografy		767
Zirconium	inorganic chemistry	546
Zoophytes		593

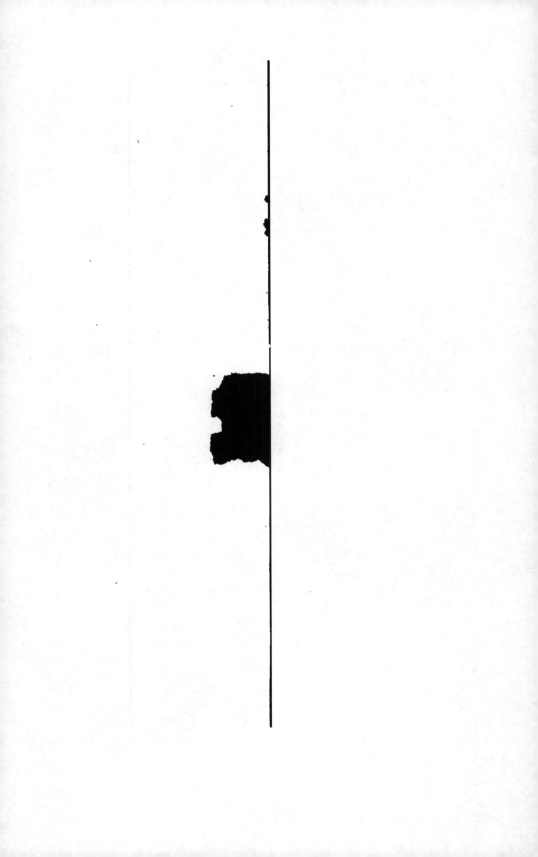

CPSIA information can be obtained
at www.ICGtesting.com
Printed in the USA
LVHW082003220622
721768LV00002B/42